THE NONUNIFORM DISCRETE FOURIER TRANSFORM AND ITS APPLICATIONS IN SIGNAL PROCESSING

THE KLUWER INTERNATIONAL SERIES
IN ENGINEERING AND COMPUTER SCIENCE

THE NONUNIFORM DISCRETE FOURIER TRANSFORM AND ITS APPLICATIONS IN SIGNAL PROCESSING

by

Sonali Bagchi
Lucent Technologies

Sanjit K. Mitra
University of California, Santa Barbara

KLUWER ACADEMIC PUBLISHERS
Boston / Dordrecht / London

Distributors for North, Central and South America:
Kluwer Academic Publishers
101 Philip Drive
Assinippi Park
Norwell, Massachusetts 02061 USA
Telephone (781) 871-6600
Fax (781) 871-6528
E-Mail <kluwer@wkap.com>

Distributors for all other countries:
Kluwer Academic Publishers Group
Distribution Centre
Post Office Box 322
3300 AH Dordrecht, THE NETHERLANDS
Telephone 31 78 6392 392
Fax 31 78 6546 474
E-Mail <orderdept@wkap.nl>

Library of Congress Cataloging-in-Publication Data

Bagchi, Sonali, 1967-
The nonuniform discrete fourier transform and its applications in signal processing / by Sonali Bagchi, Sanjit K. Mitra.
p. cm. -- (The Kluwer international series in engineering and computer science ; SECS 463)
Includes bibliographical references and index.
ISBN 0-7923-8281-1 (alk. paper)
1. Signal processing--Mathematics. 2. Fourier transformations.
I. Mitra, Sanjit Kumar. II. Title III. Series.
TK5102.9.B34 1999 98-45266
621.382'2--dc21 CIP

Printed on acid-free paper.

Printed in the United States of America

Contents

List of Figures

List of Tables

Preface

The growth in the field of digital signal processing began with the simulation of continuous-time systems in the 1950s, even though the origin of the field can be traced back to 400 years when methods were developed to solve numerically problems such as interpolation and integration. During the last 40 years, there have been phenomenal advances in the theory and application of digital signal processing.

In many applications, the representation of a discrete-time signal or a system in the frequency domain is of interest. To this end, the discrete-time Fourier transform (DTFT) and the z-transform are often used. In the case of a discrete-time signal of finite length, the most widely used frequency-domain representation is the discrete Fourier transform (DFT) which results in a finite-length sequence in the frequency domain. The DFT is simply composed of the samples of the DTFT of the sequence at equally spaced frequency points, or equivalently, the samples of its z-transform at equally spaced points on the unit circle. The DFT provides information about the spectral contents of the signal at equally spaced discrete frequency points, and thus, can be used for spectral analysis of signals. Various techniques, commonly known as the fast Fourier transform (FFT) algorithms, have been advanced for the efficient computation of the DFT. An important tool in digital signal processing is the linear convolution of two finite-length signals, which often can be implemented very efficiently using the DFT.

A generalization of the discrete Fourier transform, introduced in this book, is the nonuniform discrete Fourier transform (NDFT), which can be used to obtain frequency domain information of a finite-length signal at arbitrarily chosen frequency points. Even though the NDFT concept has been alluded to by a number of authors in recent years and applied in particular to the design of one-dimensional (1-D) and two-dimensional (2-D) finite-impulse-response (FIR) digital filters, in this book, we provide a more formal introduction to the subject and discuss a number of interesting applications including the design of 1-D and 2-D FIR digital filters. We hope that including material on the NDFT and some of its signal processing applications in a single volume will generate more interest in this topic, and lead to many other applications in the field.

This book is organized as follows. In Chapter 1, we introduce the problem of computing frequency samples of the z-transform of a finite-length sequence, and review the existing techniques. In Chapter 2, we develop the basics of the NDFT including its definition, properties and computational aspects. The NDFT is also extended to two dimensions. The ideas introduced here are utilized to develop applications of the NDFT in the following four chapters. In Chapter 3, we propose a nonuniform frequency sampling technique for designing 1-D FIR digital filters. Design examples are presented for various types of filters. In Chapter 4, we utilize the idea of the 2-D NDFT to design nonseparable 2-D FIR filters of various types. The resulting filters are compared with those designed by other existing methods. The performances of some of these filters are investigated by applying them to the decimation of digital images. In Chapter 5, we develop a design technique for synthesizing antenna patterns with nulls placed at desired angles to cancel interfering signals coming from these directions. In Chapter 6, we address the application of the NDFT in decoding dual-tone multi-frequency (DTMF) signals. We present an efficient decoding algorithm based on the subband NDFT (SB-NDFT), which achieves a fast, approximate computation of the NDFT. Concluding remarks are included in Chapter 7.

The research work reported in this book was supported by several University of California MICRO grants with matching support from Rockwell International, and Xerox Corporation. Part of the manuscript was written at the Indian Institute of Science, Bangalore during a short-term visit by one of the authors (SKM). The authors thank Professor Hrvoje Babic, University of Zagreb, Zagreb, Croatia, for his review of the manuscript of this book.

SONALI BAGCHI

SANJIT K. MITRA

1 INTRODUCTION

1.1 OVERVIEW

The representation of signals and systems in the frequency domain is an important tool in the study of signal processing, communication, and other fields. In many applications, discrete-time signals and systems are characterized by finite-length sequences. A widely used frequency-domain representation of such a sequence is its discrete Fourier transform (DFT) which corresponds to equally spaced samples of its discrete-time Fourier transform (DTFT), or equivalently, the samples of its z-transform evaluated on the unit circle in the z-plane at equally spaced points. In most signals, the energy is distributed nonuniformly in the frequency domain. Therefore, a nonuniform sampling scheme, tailored to the frequency-domain attributes of the signal, can be more useful and convenient in some applications. This observation has been the motivation for developing the concept of the *nonuniform discrete Fourier transform* (NDFT), the main subject of this book. We thus define the NDFT of a finite-length sequence as samples of its z-transform evaluated at arbitrarily chosen points in the z-plane. This concept is basically a generalization of the conventional discrete Fourier transform. In this book, we develop the basic framework of the NDFT representation, study its properties, and demonstrate its potential in signal processing by considering a variety of applications such as spectral analysis, filter design, antenna pattern synthesis, and decoding of dual-tone multi-frequency (DTMF) signals [Bagchi, 1994].

In this chapter, we first review the definition of the conventional discrete Fourier transform, and discuss some of its properties. We then discuss a number of approaches developed to compute samples of the z-transform of a finite-length sequence at specific locations in the z-plane, and point out their limitations. The approaches discussed include the chirp z-transform, the subband discrete Fourier transform, and the computation of approximate values of the magnitude of the frequency samples evaluated at nonuniformly spaced points on the unit circle.

1.2 DISCRETE FOURIER TRANSFORM

Consider a finite-length sequence $x[n]$, $0 \leq n \leq N-1$, of length N with a z-transform given by

$$X(z) = \sum_{n=0}^{N-1} x[n] z^{-n}, \tag{1.1}$$

which is a function of the complex variable z. Since $x[n]$ is a causal finite-length sequence, the region of convergence of $X(z)$ is the entire z-plane excluding the origin. The N-point *discrete Fourier transform* (DFT) $X[k]$ of the sequence $x[n]$ is defined as frequency samples $X(z_k)$, obtained by sampling $X(z)$ at N equally spaced points z_k on the unit circle, $z = e^{j\omega}$, given by

$$z_k = e^{j\frac{2\pi}{N}k}, \qquad 0 \leq k \leq N-1, \tag{1.2}$$

Thus,

$$X[k] = X(z)\big|_{z=e^{j\frac{2\pi}{N}k}} = \sum_{n=0}^{N-1} x[n] e^{-j\frac{2\pi}{N}kn}, \qquad 0 \leq k \leq N-1. \tag{1.3}$$

It should be noted that $X[k]$ is also a sequence of length N. Fig. 1.1 shows the location of the DFT samples in the z-plane for $N = 16$. Using the commonly used notation

$$W_N = e^{-j2\pi/N}, \tag{1.4}$$

the DFT expression in Eq. (1.3) can be rewritten as

$$X[k] = \sum_{n=0}^{N-1} x[n] {W_N}^{kn}, \qquad 0 \leq k \leq N-1. \tag{1.5}$$

The inverse DFT (IDFT) is given by

$$x[n] = \frac{1}{N} \sum_{k=0}^{N-1} X[k] {W_N}^{-kn}, \qquad 0 \leq n \leq N-1. \tag{1.6}$$

In matrix form, the DFT can be expressed as

$$\mathbf{X} = \mathbf{D}\mathbf{x} \tag{1.7}$$

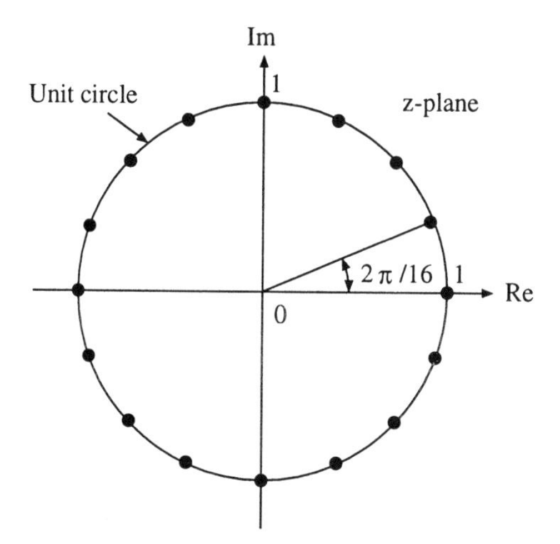

Figure 1.1. Sample locations obtained by a 16-point DFT in the z-plane.

where

$$\mathbf{X} = [X(z_0) \quad X(z_1) \quad \cdots \quad X(z_{N-1})]^{\mathbf{T}}, \tag{1.8}$$

$$\mathbf{x} = [x[0] \quad x[1] \quad \cdots \quad x[N-1]]^{\mathbf{T}}, \tag{1.9}$$

and

$$\mathbf{D} = \begin{bmatrix} 1 & 1 & \cdots & 1 \\ 1 & W_N^1 & \cdots & W_N^{N-1} \\ \vdots & \vdots & \ddots & \vdots \\ 1 & W_N^{N-1} & \cdots & W_N^{(N-1)(N-1)} \end{bmatrix}. \tag{1.10}$$

The $N \times N$ matrix $\mathbf{D}$ is called the *DFT matrix.*

The DFT has several interesting properties [Mitra, 1998] which can be utilized in many practical situations. A particularly useful application of the DFT is in computing the circular convolution and linear convolution of two finite-length sequences of equal length. For example, if $x[n]$ and $h[n]$ are two sequences of length N with $X[k]$ and $H[k]$ denoting their N-point DFTs, respectively, then the N-point IDFT $y_C[n]$ of the product $Y_C[k] = X[k]H[k]$ is given by the circular convolution of $x[n]$ and $h[n]$

$$y_C[n] = \sum_{m=0}^{N-1} x[m]h[<n-m>_N] = x[n] \, \textcircled{N} \, h[n], \qquad 0 \leq n \leq N-1, \tag{1.11}$$

where $<p>_N$ denotes p modulo N.

The linear convolution $y_L[n]$ of $x[n]$ and $h[n]$

$$y_L[n] = \sum_{m=0}^{N-1} x[m]h[n-m] = x[n] \circledast h[n], \qquad 0 \leq n \leq 2N-2, \tag{1.12}$$

can also be computed by the DFT approach. To this end, the sequences $x[n]$ and $h[n]$ are padded with $N-1$ zeros resulting in two sequences of length $2N-1$ given by

$$x_a[n] = \begin{cases} x[n], & 0 \le n \le N-1, \\ 0, & N \le n \le 2N-1, \end{cases} \tag{1.13}$$

$$h_a[n] = \begin{cases} h[n], & 0 \le n \le N-1, \\ 0, & N \le n \le 2N-1. \end{cases} \tag{1.14}$$

Let $X_a[k]$ and $H_a[k]$ denote the $(2N-1)$-point DFTs of $x_a[n]$ and $h_a[n]$, respectively. Then the $(2N-1)$-point IDFT of the product $Y_L[k] = X_a[k]H_a[k]$ is precisely the sequence $y_L[n]$.

The DFT is widely used as it can be computed very efficiently by fast Fourier transform (FFT) algorithms [Cooley and Tukey, 1965]. The number of arithmetic operations involved in the direct computation of Eq. (1.3) is $O(N^2)$. The FFT algorithms reduce this number to $O(N \log_2 N)$ if N is a power of 2, or $O(N \sum_i m_i)$ with integers m_i being prime factors of N.

1.3 CHIRP Z-TRANSFORM

Despite the computational advantages, the DFT gives only equally spaced samples on the unit circle. This implies that it has a *fixed* spectral resolution which depends only on the number of points. In an attempt to obtain a more general sampling of Eq. (1.1), Rabiner, Schafer, and Rader [Rabiner et al., 1969] introduced the *chirp z-transform* (CZT). This algorithm computes samples of the z-transform at M points which lie on circular or spiral contours beginning at any arbitrary point on the z-plane. The angular spacing of these points is an arbitrary *constant.* The sampling points are given by

$$z_k = AW^{-k}, \qquad k = 0, 1, \ldots, M-1, \tag{1.15}$$

where M is an arbitrary positive integer and both A and W are complex numbers of the form

$$A = A_0 e^{j\theta_0}$$

and

$$W = W_0 e^{j\phi_0}.$$

The locations of these sampling points in the z-plane are shown in Fig. 1.2(a). The z-plane contour defined in Eq. (1.15) starts at $z = A$ and spirals in or out with respect to the origin, depending on the value of W. The angular spacing between the samples is ϕ_0. The CZT algorithm is based on the fact that the values of the z-transform $X(z)$ given by Eq. (1.15) can be expressed as a discrete convolution of the form

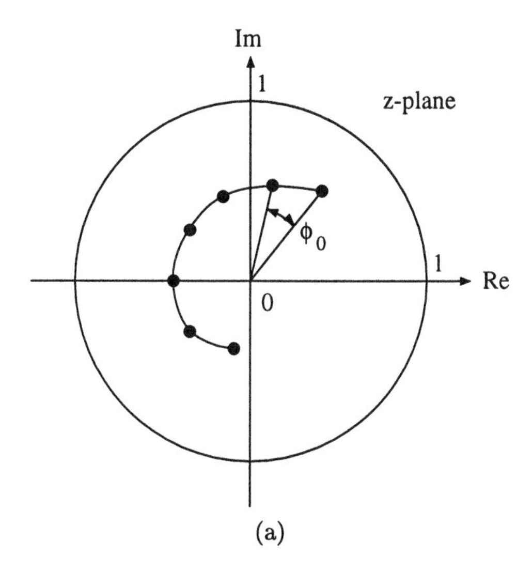

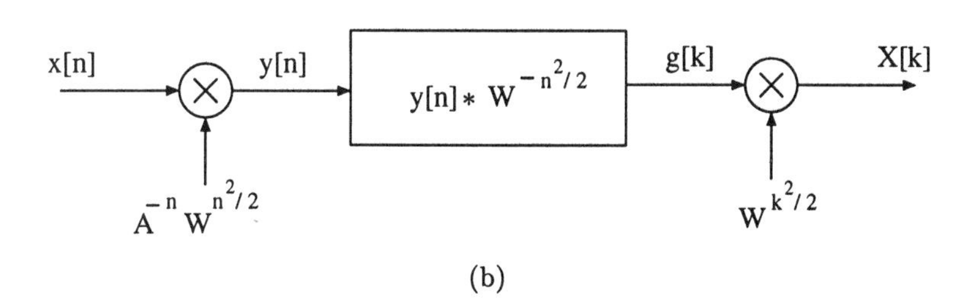

Figure 1.2. The chirp z-transform. (a) Sample locations obtained by the CZT in the z-plane. (b) Diagram illustrating the steps involved in the implementation of the CZT algorithm.

$$\begin{aligned} X(z_k) &= \sum_{n=0}^{N-1} x[n] z_k^{-n} \\ &= \sum_{n=0}^{N-1} \left(x[n] A^{-n} W^{n^2/2} \right) W^{-(k-n)^2/2} W^{k^2/2}, \qquad (1.16) \\ & \qquad 0 \leq k \leq M-1, \end{aligned}$$

as depicted in Fig. 1.2(b). High-speed convolution techniques using the FFT can be used to evaluate the CZT efficiently with $O((N+M)\log_2(N+M))$ operations, for very large values of M and N.

Thus, the CZT can be used to compute selected samples of $X(z)$ that are *equally spaced* within a particular region of the unit circle. An alternate approach is the subband DFT, discussed next.

1.4 SUBBAND DISCRETE FOURIER TRANSFORM

The dominant DFT samples in one or more portions of the frequency range can be efficiently calculated by the subband DFT computation method [Mitra et al., 1990]. The basic idea behind this method is to decompose the original sequence $x[n]$ into a sum of subsequences with frequency components in contiguous bands using a structural subband decomposition, and then to compute the DFT samples using only the subsequences containing the most energy. However, this approach results in approximate values of the DFT samples and is restricted to sequences of length N that is a power of 2.

In the first step, the sequence $x[n]$ is decomposed into two subsequences, $g_L[n]$ and $g_H[n]$, of length $N/2$ each:

$$\begin{aligned} g_L[n] &= \frac{1}{2}\{x[2n] + x[2n+1]\}, \\ g_H[n] &= \frac{1}{2}\{x[2n] - x[2n+1]\}, \qquad n = 0, 1, \ldots, N/2-1. \end{aligned} \tag{1.17}$$

These subsequences, composed of even and odd samples, are simply the low-frequency and high-frequency components of $x[n]$, downsampled by a factor of 2. The parent sequence $x[n]$ can be recovered from $g_L[n]$ and $g_H[n]$ by

$$\begin{aligned} x[2n] &= g_L[n] + g_H[n], \\ x[2n+1] &= g_L[n] - g_H[n], \qquad n = 0, 1, \ldots, N/2-1. \end{aligned} \tag{1.18}$$

If $G_L(z)$ and $G_H(z)$ denote the z-transforms of $g_L[n]$ and $g_H[n]$, then it follows from Eq. (1.18) that

$$\begin{aligned} X(z) &= \sum_{n=0}^{N-1} x[n] z^{-n} \\ &= \sum_{n=0}^{N/2-1} x[2n] z^{-2n} + \sum_{n=0}^{N/2-1} x[2n+1] z^{-(2n+1)} \\ &= \sum_{n=0}^{N/2-1} \{g_L[n] + g_H[n]\} z^{-2n} + z^{-1} \sum_{n=0}^{N/2-1} \{g_L[n] - g_H[n]\} z^{-2n} \\ &= (1+z^{-1}) \sum_{n=0}^{N/2-1} g_L[n] z^{-2n} + (1-z^{-1}) \sum_{n=0}^{N/2-1} g_H[n] z^{-2n} \\ &= (1+z^{-1})\, G_L(z^2) + (1-z^{-1})\, G_H(z^2). \end{aligned} \tag{1.19}$$

The N-point DFT $X[k]$ of $x[n]$ can then be computed by evaluating Eq. (1.19) on the unit circle, resulting in

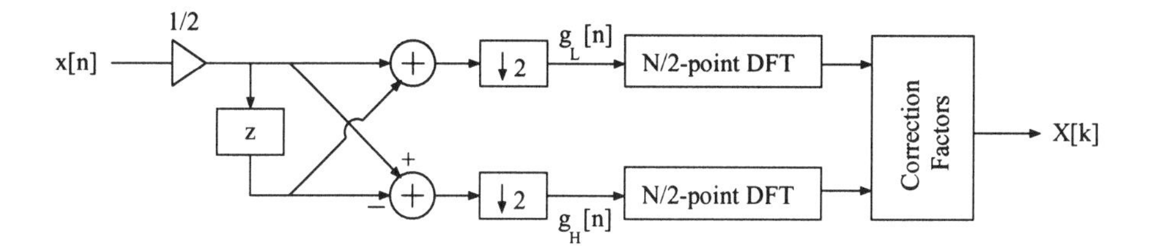

Figure 1.3. First stage in the SB-DFT computation scheme.

$$X[k] = (1 + W_N^k) G_L[\langle k \rangle_{N/2}] + (1 - W_N^k) G_H[\langle k \rangle_{N/2}], \quad 0 \leq k \leq N-1, \tag{1.20}$$

where $G_L[k]$ and $G_H[k]$ are the $(N/2)$-point DFTs of the subsequences $g_L[n]$ and $g_H[n]$, respectively. The DFT computation of Eq. (1.20) has been called the *subband DFT*, and its network interpretation is indicated in Fig. 1.3. Note that the two $(N/2)$-point DFTs can be implemented using any FFT algorithm. The subband DFT algorithm can be applied to the computation of the DFTs, $G_L[k]$ and $G_H[k]$, of the subsequences, provided $N/2$ is even, resulting in a two-stage algorithm. This process can be repeated until all DFTs are of length 2. If $N = 2^M$, the total number of complex multiplications required in a M-stage subband DFT algorithm is $N/2 \log_2 N$, which is the same as any Cooley-Tukey type algorithm. However, the total number of additions is much higher in the subband DFT algorithm.

A computationally more efficient algorithm results if the spectral separation property of the subband decomposition is taken into account. If certain subbands make negligible contributions to the energy of $x[n]$, then the DFT computation can be simplified by eliminating the calculations corresponding to these bands. For example, if $x[n]$ has most of its energy content at low frequencies, its DFT can be computed approximately by removing the contribution of $G_H[k]$ from Eq. (1.20),

$$X[k] \cong \begin{cases} (1 + W_N^k) G_L[< k >_{N/2}], & 0 \leq k < N/2 - 1, \\ 0, & N/2 \leq k \leq N-1, \end{cases} \tag{1.21}$$

assuming that $x[n]$ is a real-valued sequence. The above approximation is satisfactory in practice, since for $k \ll N$ in the low-frequency band, $|1 - W_N^k| \ll |1 + W_N^k|$, and due to the prefiltering process, $|G_H[k]| \ll |G_L[k]|$. The total computation requirement in using the approximate algorithm of Eq. (1.21) is one-half of that for the original algorithm of Eq. (1.20). In the general case of an M-stage decomposition, if L of the M bands are eliminated from calculation, the computational saving is of the order of L/M.

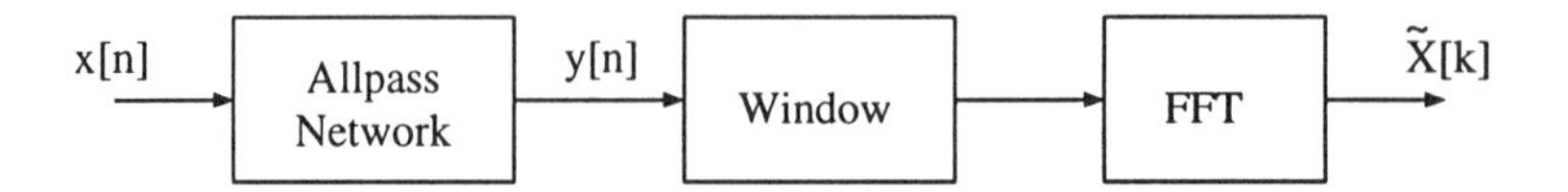

Figure 1.4. Computing the magnitude spectrum with unequal resolution using the FFT.

1.5 COMPUTATION OF NONUNIFORMLY SPACED FREQUENCY SAMPLES

To compute the magnitude spectrum with unequal resolution, Oppenheim and Johnson proposed a transformation of the original sequence $x[n]$ to a new sequence $y[n]$, so that the magnitude of the DFT $Y[k]$ of the new sequence is equal to the magnitude of frequency samples of the z-transform $X(z)$ of the original sequence at unequally spaced points on the unit circle [Oppenheim and Johnson, 1971]. Fig. 1.4 shows the basic scheme involved in their algorithm. An allpass network is used to implement a distortion of the frequency axis. Thus, the new sequence $y[n]$ is of infinite length and has to be windowed by a finite-length window before the FFT can be computed. Since the windowing causes a smearing of the spectrum before the spectral samples can be computed, the method yields only approximate results.

1.6 SUMMARY

Several different frequency-domain representations of a finite-length sequence have been reviewed in this chapter. All such representations consist basically of a finite number of samples of the z-transform of the sequence at selected points in the z-plane. The first representation considered is the popular discrete Fourier transform (DFT) which is composed of samples of the z-transform at equally spaced points on the unit circle, with the first sample located at $z = 1$. An interesting application of the DFT is the computation of the circular convolution of two finite-length sequences by computing the inverse DFT of the product of their DFTs. By padding the two sequences with an appropriate number of zero-valued samples, this approach can be used to compute the linear convolution of two finite-length sequences.

The second frequency-domain representation discussed is the chirp z-transform, composed of samples of the z-transform at equally spaced points on an arc in the z-plane, beginning at an arbitrary point. The chirp z-transform can be expressed as a linear convolution by simple algebraic manipulation, and can then be computed efficiently using the DFT-based approach.

In applications requiring computation of approximate values of DFT samples at equally spaced points over a specified portion of the unit circle, an efficient algorithm is the subband DFT, discussed next. The last algorithm discussed in this chapter involves computation of samples of the z-transform at unequally spaced points on the unit circle. This approach provides approximate values of the magnitudes of the desired samples.

To obtain a more general sampling of the z-transform of a finite-length sequence, we introduce the concept of the *nonuniform discrete Fourier transform* (NDFT). The NDFT represents samples of the z-transform at distinct points located anywhere in the z-plane. In the following chapter, we define the NDFT and the inverse NDFT, discuss several methods for their computation, describe the properties of the NDFT, present the subband NDFT, and extend the NDFT to two dimensions.

2 THE NONUNIFORM DISCRETE FOURIER TRANSFORM

We develop the concept of the nonuniform discrete Fourier transform (NDFT) in this chapter. The basic idea is introduced in Section 2.1. We investigate the properties of the NDFT in Section 2.2. Various methods for computing the NDFT are discussed in Section 2.3. We outline the subband NDFT in Section 2.4. Finally, the concept of the NDFT is extended to two dimensions in Section 2.5.

2.1 BASIC CONCEPTS

We define the NDFT at the beginning of this section. An example of spectral analysis is presented to demonstrate how the variable spectral resolution offered by the NDFT can be utilized. Then, we show that the computation of the inverse NDFT is equivalent to polynomial interpolation.

2.1.1 Definition

The *nonuniform discrete Fourier transform* (NDFT) of a sequence $x[n]$ of length N is defined as [Mitra et al., 1992; Bagchi and Mitra, 1996a]

$$X(z_k) = \sum_{n=0}^{N-1} x[n] z_k^{-n}, \qquad k = 0, 1, \ldots, N-1, \tag{2.1}$$

where $z_0, z_1, \ldots, z_{N-1}$ are distinct points located arbitrarily in the z-plane. We can express Eq. (2.1) in a matrix form as

$$\mathbf{X} = \mathbf{D}\mathbf{x} \tag{2.2}$$

where

$$\mathbf{X} = \begin{bmatrix} X(z_0) \\ X(z_1) \\ \vdots \\ X(z_{N-1}) \end{bmatrix}, \qquad \mathbf{x} = \begin{bmatrix} x[0] \\ x[1] \\ \vdots \\ x[N-1] \end{bmatrix}, \tag{2.3}$$

and

$$\mathbf{D} = \begin{bmatrix} 1 & z_0^{-1} & z_0^{-2} & \cdots & z_0^{-(N-1)} \\ 1 & z_1^{-1} & z_1^{-2} & \cdots & z_1^{-(N-1)} \\ \vdots & \vdots & \vdots & \ddots & \vdots \\ 1 & z_{N-1}^{-1} & z_{N-1}^{-2} & \cdots & z_{N-1}^{-(N-1)} \end{bmatrix}. \tag{2.4}$$

Note that the NDFT matrix $\mathbf{D}$ is fully specified by the choice of the N points z_k. A matrix of this form is known as a Vandermonde matrix [Golub and Loan, 1983]. It can be shown that the determinant of $\mathbf{D}$ can be expressed in a factored form as [Davis, 1975]

$$\det(\mathbf{D}) = \prod_{i \neq j,\ i > j} (z_i{}^{-1} - z_j{}^{-1}). \tag{2.5}$$

Consequently, $\mathbf{D}$ is nonsingular provided the N sampling points, $z_0, z_1, \ldots, z_{N-1}$, are *distinct*. Thus, the inverse NDFT exists and is unique, given by

$$\mathbf{x} = \mathbf{D}^{-1}\mathbf{X}. \tag{2.6}$$

As a *special case*, consider the situation when the points z_k are located at equally spaced angles on the unit circle in the z-plane. This corresponds to the conventional discrete Fourier transform (DFT). The matrix $\mathbf{D}$ then reduces to the unitary DFT matrix shown in Eq. (1.10).

Example 2.1

Consider the linear convolution of two length-2 sequences,

$$\begin{aligned} \{a[n]\} &= \{\, a_0 \quad a_1 \,\}, \\ \{b[n]\} &= \{\, b_0 \quad b_1 \,\}, \end{aligned}$$

resulting in a length-3 sequence

$$\{h[n]\} = \{\, h_0 \quad h_1 \quad h_2 \,\},$$

where

$$\begin{aligned}
h_0 &= a_0 b_0, \\
h_1 &= a_0 b_1 + a_1 b_0, \\
h_2 &= a_1 b_1.
\end{aligned}$$

Thus, a direct implementation of the convolution requires 4 multiplications.

The convolution can also be implemented in the z-domain as a product of two polynomials,

$$\begin{aligned}
A(z) &= a_0 + a_1 z^{-1}, \\
B(z) &= b_0 + b_1 z^{-1},
\end{aligned}$$

resulting in a polynomial

$$\begin{aligned}
H(z) &= h_0 + h_1 z^{-1} + h_2 z^{-2} \\
&= A(z)B(z) = (a_0 + a_1 z^{-1})(b_0 + b_1 z^{-1}).
\end{aligned}$$

A computationally efficient method to determine the coefficients of $H(z)$ is through the NDFT, as illustrated next. We evaluate the NDFT samples of $H(z) = A(z)B(z)$ at three distinct points in the z-plane, $z_0 = \infty, z_1 = 1$, and $z_2 = -1$, resulting in

$$\begin{aligned}
H(z_0) &= H(\infty) = A(\infty)B(\infty) = a_0 b_0, \\
H(z_1) &= H(1) = A(1)B(1) = (a_0 + a_1)(b_0 + b_1), \\
H(z_2) &= H(-1) = A(-1)B(-1) = (a_0 - a_1)(b_0 - b_1).
\end{aligned}$$

Note that the computation of the NDFT samples requires only 3 multiplications. From Eq. (2.4), the 3×3 NDFT matrix for this case is given by

$$\mathbf{D} = \begin{bmatrix} 1 & z_0^{-1} & z_0^{-2} \\ 1 & z_1^{-1} & z_1^{-2} \\ 1 & z_2^{-1} & z_2^{-2} \end{bmatrix} = \begin{bmatrix} 1 & 0 & 0 \\ 1 & 1 & 1 \\ 1 & -1 & 1 \end{bmatrix},$$

and has an inverse

$$\mathbf{D}^{-1} = \begin{bmatrix} 1 & 0 & 0 \\ 0 & \frac{1}{2} & -\frac{1}{2} \\ -1 & \frac{1}{2} & \frac{1}{2} \end{bmatrix}.$$

Hence the 3-point NDFT samples are related to the sequence $\{h[n]\}$ through

$$\begin{bmatrix} H(\infty) \\ H(1) \\ H(-1) \end{bmatrix} = \mathbf{D} \begin{bmatrix} h_0 \\ h_1 \\ h_2 \end{bmatrix} = \begin{bmatrix} 1 & 0 & 0 \\ 1 & 1 & 1 \\ 1 & -1 & 1 \end{bmatrix} \begin{bmatrix} h_0 \\ h_1 \\ h_2 \end{bmatrix}.$$

or equivalently,

$$\begin{aligned}
h_0 &= H(\infty) = a_0 b_0, \\
h_1 &= \frac{1}{2}H(1) - \frac{1}{2}H(-1) \\
&= \frac{1}{2}(a_0 + a_1)(b_0 + b_1) - \frac{1}{2}(a_0 - a_1)(b_0 - b_1) \\
&= a_0 b_1 + a_1 b_0, \\
h_2 &= -H(\infty) + \frac{1}{2}H(1) + \frac{1}{2}H(-1) \\
&= -a_0 b_0 + \frac{1}{2}(a_0 + a_1)(b_0 + b_1) - \frac{1}{2}(a_0 - a_1)(b_0 - b_1) \\
&= a_1 b_1.
\end{aligned}$$

It should be noted that the above method is precisely the Cook-Tom algorithm for fast convolution of two length-2 sequences [Blahut, 1985].

2.1.2 Spectral Analysis Using the NDFT

The DFT has been widely used in analyzing the frequency content of signals. Typically, a continuous-time signal is sampled, a finite-duration window is applied to it to obtain a sequence of length N, and then an R-point DFT, with $R \geq N$, is used to compute R samples of the spectrum of the windowed sequence. Often, misleading results may be obtained using this procedure if a sufficient number of points are not taken. For example, if the signal contains two *closely spaced sinusoids*, the resulting DFT may fail to resolve the corresponding peaks in the spectrum, even when the window has not smeared out the peaks.

For fixed-length transforms, with some knowledge of the frequency-content of the signal, we can improve the resolution by using the NDFT to compute the spectral samples. Note that we are referring to the resolution involved in sampling the windowed spectrum, which is different from the resolution involved in the windowing operation itself (the latter depends on the mainlobe width of the window). A simple example to illustrate our idea is shown in Fig. 2.1. The discrete-time Fourier transform of a windowed sequence is shown in Fig. 2.1(a). Fig. 2.1(b) shows the spectrum obtained by using a 64-point DFT. From these DFT samples, we might *erroneously* conclude that there is only one major tone present in the signal. As shown in Fig. 2.1(c), the unsatisfactory resolution is *improved* by using a 64-point NDFT, where the spectrum is sampled more densely in the critical region. In this particular example, the density of samples is simply doubled in the low-frequency band, and halved in the high-frequency band, so that the total number of samples remains the same.

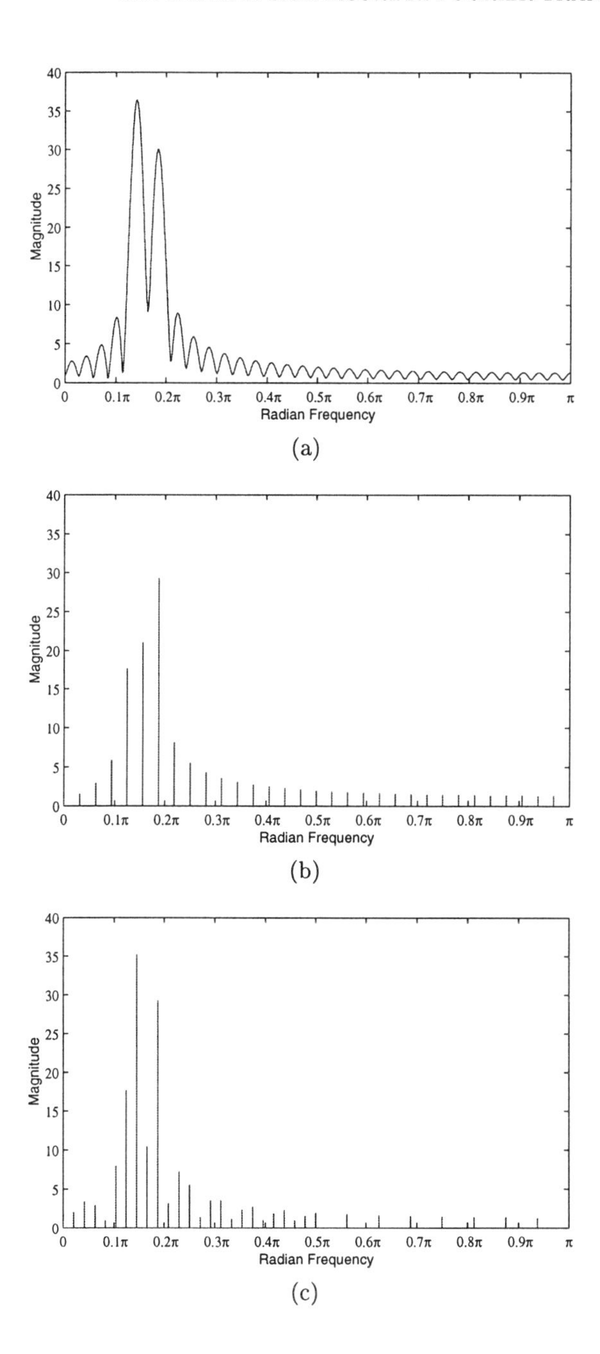

Figure 2.1. Spectral analysis example. (a) Original spectrum composed of two tones whose frequencies are to be estimated. (b) Estimate obtained by a 64-point DFT. (c) Estimate obtained by a 64-point NDFT with dense sampling in the low frequency band.

2.1.3 Computation of the Inverse NDFT

The problem of computing the inverse NDFT, i.e., determining $\mathbf{x}$ from a given NDFT vector $\mathbf{X}$, is equivalent to solving the Vandermonde system in Eq. (2.6). This can be stated alternatively as a polynomial interpolation problem, which arises in many areas of engineering and numerical analysis. The basis for the existence of the solution is given by the simple theorem of polynomial interpolation. This theorem is restated here for convenience [Davis, 1975]: Given N distinct (real or complex) points, $z_0, z_1, \ldots, z_{N-1}$, and N (real or complex) values, $f_0, f_1, ..., f_{N-1}$, there exists a unique polynomial,

$$P(z) = a_0 + a_1 z + a_2 z^2 + \cdots + a_{N-1} z^{N-1},$$

for which

$$P(z_k) = f_k, \qquad k = 0, 1, \ldots, N-1.$$

Although the solution to this problem is unique, *various methods* of solving the problem exist. This leads to different ways of computing the inverse NDFT:

(1) Direct Method

Given the NDFT $\mathbf{X}$ and the NDFT matrix $\mathbf{D}$, the inverse NDFT $\mathbf{x}$ is found directly by solving the linear system in Eq. (2.2), using Gaussian elimination. This involves $O(N^3)$ operations.

The same problem can be solved more *efficiently* by using polynomial interpolation. In this approach, the z-transform $X(z)$ is directly determined in terms of the NDFT coefficients,

$$\hat{X}[k] = X(z_k), \qquad k = 0, 1, \ldots, N-1, \tag{2.7}$$

by using polynomial interpolation methods. The inverse NDFT $x[n]$ can then be identified as the coefficients of this interpolating polynomial. This approach is used in two existing interpolation methods, Lagrange and Newton interpolation.

(2) Lagrange Interpolation

$X(z)$ is expressed as the Lagrange polynomial of order $N-1$,

$$X(z) = \sum_{k=0}^{N-1} \frac{L_k(z)}{L_k(z_k)} \hat{X}[k], \tag{2.8}$$

where $L_0(z), L_1(z), \ldots, L_{N-1}(z)$ are the fundamental polynomials, defined as

$$L_k(z) = \prod_{i \neq k} (1 - z_i z^{-1}), \qquad k = 0, 1, \ldots, N-1. \tag{2.9}$$

Example 2.2

We consider the implementation of the convolution in Example 2.1 using the Lagrange interpolation method. Here, we express $H(z)$ as

$$H(z) = \frac{L_0(z)}{L_0(z_0)}H(z_0) + \frac{L_1(z)}{L_1(z_1)}H(z_1) + \frac{L_2(z)}{L_2(z_2)}H(z_2),$$

where

$$\begin{aligned} L_0(z) &= (1 - z_1 z^{-1})(1 - z_2 z^{-1}), \\ L_1(z) &= (1 - z_0 z^{-1})(1 - z_2 z^{-1}), \\ L_2(z) &= (1 - z_0 z^{-1})(1 - z_1 z^{-1}). \end{aligned}$$

With $z_0 = \infty$, $z_1 = 1$, and $z_2 = -1$, we get

$$\begin{aligned} L_0(z) &= (1 - z^{-1})(1 + z^{-1}), \\ L_1(z) &= -z_0 z^{-1}(1 + z^{-1}), \\ L_2(z) &= -z_0 z^{-1}(1 - z^{-1}), \end{aligned}$$

and

$$\begin{aligned} L_0(z_0) &= L_0(\infty) = 1, \\ L_1(z_1) &= L_1(1) = -2z_0, \\ L_2(z_2) &= L_2(-1) = 2z_0. \end{aligned}$$

Therefore,

$$\begin{aligned} H(z) &= (1 - z^{-2})H(z_0) + \frac{z_0 z^{-1}(1 + z^{-1})}{2z_0}H(z_1) - \frac{z_0 z^{-1}(1 - z^{-1})}{2z_0}H(z_2) \\ &= (1 - z^{-2})H(\infty) + \frac{1}{2}z^{-1}(1 + z^{-1})H(1) - \frac{1}{2}z^{-1}(1 - z^{-1})H(-1) \\ &= (1 - z^{-2})a_0 b_0 + \frac{1}{2}z^{-1}(1 + z^{-1})(a_0 + a_1)(b_0 + b_1) \\ &\quad -\frac{1}{2}z^{-1}(1 - z^{-1})(a_0 - a_1)(b_0 - b_1) \\ &= a_0 b_0 + (a_1 b_0 + a_0 b_1)z^{-1} + a_2 b_2 z^{-2}. \end{aligned}$$

(3) Newton Interpolation

$X(z)$ is expressed in the form

$$X(z) = c_0 + c_1(1 - z_0 z^{-1}) + c_2(1 - z_0 z^{-1})(1 - z_1 z^{-1}) + \cdots + c_{N-1}\prod_{k=0}^{N-2}(1 - z_k z^{-1}), \tag{2.10}$$

where the coefficient c_j is called the divided difference of the jth order of $\hat{X}[0], \hat{X}[1], \ldots, \hat{X}[j]$ with respect to $z_0, z_1, \ldots, z_j$. The divided differences are computed recursively as follows:

$$\begin{aligned} c_0 &= \hat{X}[0] \\ c_1 &= \frac{\hat{X}[1] - c_0}{1 - z_0 z_1^{-1}} \\ c_2 &= \frac{\hat{X}[2] - c_0 - c_1(1 - z_0 z_2^{-1})}{(1 - z_0 z_2^{-1})(1 - z_1 z_2^{-1})} \\ &\vdots \end{aligned} \tag{2.11}$$

Note that each c_j is a linear combination of the $\hat{X}[k]$, and moreover, c_j depends only on $\hat{X}[0], \hat{X}[1], \ldots, \hat{X}[j]$, and $z_0, z_1, \ldots, z_j$. In the Lagrange representation, if we include an additional point and increase the order of the interpolating polynomial, all the fundamental polynomials change and, consequently, have to be recomputed. In the Newton representation, this can be accomplished by simply adding one more term. Thus, it has a *permanence property* which is a characteristic of the Fourier series and other orthogonal and biorthogonal expansions.

Since the coefficients c_j are computed by solving the lower triangular system of equations,

$$\mathbf{Lc} = \mathbf{X}, \tag{2.12}$$

where

$$\mathbf{L} = \begin{bmatrix} 1 & 0 & 0 & \cdots & 0 \\ 1 & (1 - z_0 z_1^{-1}) & 0 & \cdots & 0 \\ 1 & (1 - z_0 z_2^{-1}) & 0 & \cdots & 0 \\ \vdots & \vdots & \vdots & \ddots & \vdots \\ 1 & (1 - z_0 z_{N-1}^{-1}) & \cdots & \prod_{k=0}^{N-2} (1 - z_k z_{N-1}^{-1}) & \end{bmatrix}, \tag{2.13}$$

and

$$\mathbf{c} = \begin{bmatrix} c_0 \\ c_1 \\ c_2 \\ \vdots \\ c_{N-1} \end{bmatrix}, \qquad \mathbf{X} = \begin{bmatrix} \hat{X}[0] \\ \hat{X}[1] \\ \hat{X}[2] \\ \vdots \\ \hat{X}[N-1] \end{bmatrix}, \tag{2.14}$$

this involves $O(N^2)$ operations. The sequence $x[n]$ can now be easily computed from the c_j. Thus, Newton interpolation provides a *more efficient* method of solving for the inverse NDFT.

Example 2.3

In this example, we solve the convolution problem of Example 2.1 using the Newton interpolation method. Here,

$$H(z) = c_0 + c_1(1 - z_0 z^{-1}) + c_2(1 - z_0 z^{-1})(1 - z_1 z^{-1}),$$

where

$$\begin{aligned} c_0 &= H(z_0), \\ c_1 &= \frac{H(z_1) - c_0}{1 - z_0 {z_1}^{-1}}, \\ c_2 &= \frac{H(z_2) - c_0 - c_1(1 - z_0 {z_2}^{-1})}{(1 - z_0 {z_2}^{-1})(1 - z_1 {z_2}^{-1})}. \end{aligned}$$

Substituting $z_0 = \infty$, $z_1 = 1$, and $z_2 = -1$, we get three samples of $H(z)$,

$$\begin{aligned} H(z_0) &= a_0 b_0, \\ H(z_1) &= (a_0 + a_1)(b_0 + b_1), \\ H(z_2) &= (a_0 - a_1)(b_0 - b_1). \end{aligned}$$

Using these samples, we get

$$\begin{aligned} c_0 &= a_0 b_0, \\ c_1 &= \frac{(a_0 + a_1)(b_0 + b_1) - a_0 b_0}{1 - z_0 {z_1}^{-1}} = \frac{a_1 b_1 + a_0 b_1 + a_1 b_0}{1 - z_0}, \\ c_2 &= \frac{(a_0 - a_1)(b_0 - b_1) - a_0 b_0 - \frac{(a_1 b_1 + a_0 b_1 + a_1 b_0)(1 + z_0)}{1 - z_0}}{2(1 + z_0)} \\ &= \frac{(a_1 b_1 - a_0 b_1 - a_1 b_0) + (a_1 b_1 + a_0 b_1 + a_1 b_0)}{2(1 + z_0)} = \frac{a_1 b_1}{1 + z_0}. \end{aligned}$$

Therefore,

$$\begin{aligned} H(z) &= a_0 b_0 + \left(\frac{a_1 b_1 + a_0 b_1 + a_1 b_0}{1 - z_0} \right) (1 - z_0 {z_1}^{-1}) \\ &\quad + \frac{a_1 b_1}{1 + z_0} (1 - z_0 z^{-1})(1 - z^{-1}) \\ &= a_0 b_0 + (a_1 b_1 + a_0 b_1 + a_1 b_0) z^{-1} - a_1 b_1 z^{-1}(1 - z^{-1}) \\ &= a_0 b_0 + (a_0 b_1 + a_1 b_0) z^{-1} + a_1 b_1 z^{-2}. \end{aligned}$$

2.2 PROPERTIES OF THE NDFT

We study some relevant properties of the NDFT in this section. Many of these properties are analogous to properties of the z-transform and Fourier transform. However, unlike the DFT, there is no implicit periodicity in the general NDFT representation of finite-length sequences. As a result, some properties of the DFT such as circular shift and duality do not hold for the NDFT.

2.2.1 Basic Properties

Let $X(z_k)$ and $Y(z_k)$ denote the NDFT of the sequences $x[n]$ and $y[n]$, respectively. For notational convenience, we denote an NDFT pair as

$$x[n] \overset{NDFT}{\longleftrightarrow} X(z_k)$$

The basic properties are summarized in Table 2.1. Proofs of these properties follow.

Property 1 *Linearity*

$$\alpha x[n] + \beta y[n] \overset{NDFT}{\longleftrightarrow} \alpha X(z_k) + \beta Y(z_k) \tag{2.15}$$

If $x[n]$ has length N_1 and $y[n]$ has length N_2, then the linearly combined sequence

$$w[n] = \alpha x[n] + \beta y[n]$$

will have a maximum length of $N_3 = \max(N_1, N_2)$. Thus, the NDFTs must be computed with $N \geq N_3$. For example, if $N_1 < N_2$, $X(z_k)$ is the N_2-point NDFT of the sequence $x[n]$ padded with $N_2 - N_1$ zeros. Of course, NDFTs of greater length can be computed by padding both sequences with zeros.

Property 2 *Shifting in the time domain*

Suppose $x[n]$ is zero outside the range $0 \leq n \leq N-1$. Then,

$$y[m] \overset{NDFT}{\longleftrightarrow} z_k^{-n_0} X(z_k) \tag{2.16}$$

where n_0 is an integer, and

$$y[m] = x[m - n_0], \qquad m = n_0, n_0 + 1, \ldots, n_0 + N - 1.$$

The NDFT of $y[m]$ is given by

$$Y(z_k) = \sum_{m=n_0}^{n_0+N-1} x[m - n_0] z_k^{-m}. \tag{2.17}$$

Substituting $n = m - n_0$ in the above equation, we obtain

$$Y(z_k) = \sum_{n=0}^{N-1} x[n] z_k^{-n-n_0} = z_k^{-n_0} X(z_k). \tag{2.18}$$

Thus, a shift in the time domain corresponds to the multiplication of the NDFT $X(z_k)$ by a complex factor of $z_k^{-n_0}$.

Property 3 *Scaling and shifting in the z-domain*

$$z_s^n x[n] \overset{NDFT}{\longleftrightarrow} X(z_k/z_s) \tag{2.19}$$

Table 2.1. Basic properties of the NDFT.

Property 1 **Linearity**

$$\alpha x[n] + \beta y[n] \overset{NDFT}{\longleftrightarrow} \alpha X(z_k) + \beta Y(z_k)$$

Property 2 **Shifting in the time domain**

$$x[m - n_0] \overset{NDFT}{\longleftrightarrow} z_k^{-n_0} X(z_k)$$
$$(m = n_0, n_0 + 1, \ldots, n_0 + N - 1)$$

Property 3 **Scaling and shifting in the z-domain**

$$z_s^n x[n] \overset{NDFT}{\longleftrightarrow} X(z_k/z_s)$$

Property 4 **Time reversal**

$$x[-m] \overset{NDFT}{\longleftrightarrow} X(1/z_k)$$

For real $x[n]$,

$$x[-m] \overset{NDFT}{\longleftrightarrow} X^*(z_k)$$
$$(m = -N + 1, -N + 2, \ldots, 1, 0)$$

Property 5 **Conjugation**

$$x^*[n] \overset{NDFT}{\longleftrightarrow} X^*(z_k^*)$$

Property 6 **Real part of sequence**

$$Re\{x[n]\} \overset{NDFT}{\longleftrightarrow} \tfrac{1}{2}\{X(z_k) + X^*(z_k^*)\}$$

Property 7 **Imaginary part of sequence**

$$j\, Im\{x[n]\} \overset{NDFT}{\longleftrightarrow} \tfrac{1}{2}\{X(z_k) - X^*(z_k^*)\}$$

Property 8 **Symmetry properties**

For real $x[n]$,

$$X(z_k) = X^*(z_k^*)$$
$$Re\,\{X(z_k)\} = Re\,\{X(z_k^*)\}$$
$$Im\,\{X(z_k)\} = -\,Im\,\{X(z_k^*)\}$$
$$|X(z_k)| = |X(z_k^*)|$$
$$arg\,\{X(z_k)\} = -\,arg\,\{X(z_k^*)\}$$

Consider the sequence,

$$y[n] = z_s^n x[n], \qquad n = 0, 1, \ldots, N - 1.$$

The NDFT of $y[n]$ is given by

$$Y(z_k) = \sum_{n=0}^{N-1} z_s^n x[n] z_k^{-n} = \sum_{n=0}^{N-1} x[n](z_k/z_s)^{-n} = X(z_k/z_s). \tag{2.20}$$

To interpret the meaning of the above scaling, let

$$z_s = \alpha e^{j\phi}$$

and

$$z_k = r_k e^{j\omega_k}, \qquad k = 0, 1, \ldots, N-1,$$

where α and r_k are non-negative. Then, we can express the NDFT $Y(z_k)$ as

$$Y(z_k) = X(z_k{}'), \tag{2.21}$$

where

$$z_k{}' = \frac{z_k}{z_s} = \frac{r_k}{\alpha} e^{j(\omega_k - \phi)}.$$

Therefore, as a result of multiplying the original sequence by an exponential sequence z_s^n, the sample locations of the NDFT in the z-plane undergo a scaling in radius and a shift in angle. As a special case, let $\alpha = 1$. This corresponds to a shift in the angles of the sample locations only, and is equivalent to a frequency shift or translation, associated with modulation in the time domain by a complex exponential sequence $e^{j\phi n}$.

Property 4 *Time reversal*

$$y[m] \overset{NDFT}{\longleftrightarrow} X(1/z_k) \tag{2.22}$$

where

$$y[m] = x[-m], \qquad m = -N+1, -N+2, \ldots, 1, 0,$$

and $x[n]$ is zero outside the range $0 \leq n \leq N-1$. The NDFT of $y[m]$ is given by

$$Y(z_k) = \sum_{m=-N+1}^{0} x[-m] z_k^{-m}. \tag{2.23}$$

Substituting $n = -m$ in the above, we obtain

$$Y(z_k) = \sum_{n=0}^{N-1} x[n] z_k^n = X(1/z_k). \tag{2.24}$$

Thus, time reversal of the sequence results in the inversion of the sample locations in the z-plane.

As a special case, if $x[n]$ is real and $z_k = e^{j\omega_k}$, then we have

$$X(1/z_k) = X(z_k^*). \tag{2.25}$$

Since $x[n]$ is real, we can apply the result of Property 8(i) shown in Eq. (2.33) (see page 24), to obtain

$$X(1/z_k) = X^*(z_k).$$

Therefore, we get the NDFT pair

$$y[m] \overset{NDFT}{\longleftrightarrow} X^*(z_k). \tag{2.26}$$

In this case, time reversal of the sequence simply corresponds to a conjugation of the NDFT.

Property 5 *Conjugation*

$$x^*[n] \overset{NDFT}{\longleftrightarrow} X^*(z_k^*) \tag{2.27}$$

Let

$$y[n] = x^*[n], \qquad n = 0, 1, \ldots, N-1.$$

The NDFT of $y[n]$ is given by

$$Y(z_k) = \sum_{n=0}^{N-1} x^*[n] z_k^{-n} = \left\{ \sum_{n=0}^{N-1} x[n](z_k^*)^{-n} \right\}^* = X^*(z_k^*). \tag{2.28}$$

Thus, the conjugation of the original sequence leads to the conjugation of its NDFT, which has been evaluated at the conjugate sample locations in the z-plane.

Property 6 *Real part of sequence*

$$Re\{x[n]\} \overset{NDFT}{\longleftrightarrow} \frac{1}{2}\{X(z_k) + X^*(z_k^*)\} \tag{2.29}$$

The real part of $x[n]$ can be expressed as

$$p[n] = \frac{1}{2}\{x[n] + x^*[n]\}.$$

The NDFT of $p[n]$ can be obtained by applying the result of Property 5 shown in Eq. (2.27), as

$$P(z_k) = \frac{1}{2}\{X(z_k) + X^*(z_k^*)\}. \tag{2.30}$$

Property 7 *Imaginary part of sequence*

$$j\,Im\{x[n]\} \overset{NDFT}{\longleftrightarrow} \frac{1}{2}\{X(z_k) - X^*(z_k^*)\} \tag{2.31}$$

The imaginary part of $x[n]$ can be expressed as

$$q[n] = \frac{1}{2j}\{x[n] - x^*[n]\}.$$

The NDFT of $q[n]$ can be obtained by applying the result of Property 5 shown in Eq. (2.27), so that

$$jQ(z_k) = \frac{1}{2}\{X(z_k) - X^*(z_k^*)\}. \tag{2.32}$$

Property 8 *Symmetry properties*

The following properties hold when $x[n]$ is real.

(i)

$$X(z_k) = X^*(z_k^*) \tag{2.33}$$

Since $x[n]$ is real,

$$x[n] = x^*[n]. \tag{2.34}$$

By applying the result of Property 5 shown in Eq. (2.27), we get the desired result.

(ii)

$$Re\{X(z_k)\} = Re\{X(z_k^*)\} \tag{2.35}$$

Let us express $X(z_k)$ as a complex number,

$$X(z_k) = g[k] + jh[k], \qquad k = 0, 1, \ldots, N-1, \tag{2.36}$$

where

$$g[k] = Re\{X(z_k)\}, \qquad h[k] = Im\{X(z_k)\}.$$

Conjugating both sides of Eq. (2.33), we have

$$X^*(z_k) = X(z_k^*). \tag{2.37}$$

But, from Eq. (2.36),

$$X^*(z_k) = g[k] - jh[k]. \tag{2.38}$$

Comparing Eqs. (2.36) and (2.38), we obtain

$$g[k] = Re\{X(z_k)\} = Re\{X(z_k^*)\}. \tag{2.39}$$

(iii)

$$Im\{X(z_k)\} = -\,Im\{X(z_k^*)\} \tag{2.40}$$

Comparing Eqs. (2.36) and (2.38), we obtain

$$h[k] = Im\{X(z_k)\} = -\,Im\{X(z_k^*)\}. \tag{2.41}$$

(iv)

$$|X(z_k)| = |X(z_k^*)| \tag{2.42}$$

This follows directly from Eq. (2.37).

(v)

$$arg\{X(z_k)\} = -\,arg\{X(z_k^*)\} \tag{2.43}$$

This also follows from Eq. (2.37).

2.2.2 Linear Convolution Using the NDFT

We now show that the linear convolution of two sequences can be computed using the NDFT. Let $y_L[n]$ be the linear convolution of two N-point sequences, $x[n]$ and $h[n]$. Using vector notation, this convolution can be expressed as

$$\mathbf{y_L} = \mathbf{x} \circledast \mathbf{h}, \tag{2.44}$$

where

$$\mathbf{x} = [\, x[0] \quad x[1] \quad \cdots \quad x[N-1]\,]^{\mathbf{T}}, \tag{2.45}$$

$$\mathbf{h} = [\, h[0] \quad h[1] \quad \cdots \quad h[N-1]\,]^{\mathbf{T}}, \tag{2.46}$$

and

$$\mathbf{y_L} = [\, y[0] \quad y[1] \quad \cdots \quad y[2N-2]\,]^{\mathbf{T}}. \tag{2.47}$$

Let $\mathbf{X}$ and $\mathbf{H}$ denote the N-point NDFTs of $\mathbf{x}$ and $\mathbf{h}$ respectively, i.e.,

$$\mathbf{X} = \mathbf{D}\mathbf{x}, \qquad \mathbf{H} = \mathbf{D}\mathbf{h}, \tag{2.48}$$

where $\mathbf{D}$ is an $N \times N$ NDFT matrix with sampling points $z_0, z_1, \ldots, z_{N-1}$, as given in Eq. (2.4).

Now, if we multiply the NDFTs $\mathbf{X}$ and $\mathbf{H}$ (componentwise), and take the inverse NDFT of the product, we obtain an N-point sequence $\mathbf{y_C}$,

$$\mathbf{y_C} = [\, y_C[0] \quad y_C[1] \quad \cdots \quad y_C[N-1]\,]^{\mathbf{T}}, \tag{2.49}$$

given by

$$\mathbf{y_C} = \mathbf{D^{-1}}[\mathbf{X} \cdot \mathbf{H}]. \tag{2.50}$$

Here, the dot "$\cdot$" denotes the componentwise product of the two vectors. Note that in the special case of the DFT, $\mathbf{y_C}$ corresponds to the circular convolution of the sequences $\mathbf{x}$ and $\mathbf{h}$, where

$$y_C[n] = x[n] \, \textcircled{N} \, h[n], \qquad 0 \leq n \leq N-1. \tag{2.51}$$

In order to find the relationship between $\mathbf{y_C}$ and $\mathbf{y}$, we compute the z-transform of both sides of Eq. (2.44) to obtain

$$X(z)H(z) = Y_L(z). \tag{2.52}$$

Evaluating both sides of Eq. (2.52) at the points $z = z_0, z_1, \ldots, z_{N-1}$, and expressing the result in matrix form, we have

$$\mathbf{X} \cdot \mathbf{H} = \begin{bmatrix} 1 & z_0^{-1} & z_0^{-2} & \cdots & z_0^{-(N-1)} & \cdots & z_0^{-(2N-2)} \\ 1 & z_1^{-1} & z_1^{-2} & \cdots & z_1^{-(N-1)} & \cdots & z_1^{-(2N-2)} \\ \vdots & \vdots & \vdots & \ddots & \vdots & \ddots & \vdots \\ 1 & z_{N-1}^{-1} & z_{N-1}^{-2} & \cdots & z_{N-1}^{-(N-1)} & \cdots & z_{N-1}^{-(2N-2)} \end{bmatrix} \begin{bmatrix} y[0] \\ y[1] \\ \vdots \\ y[2N-2] \end{bmatrix}, \tag{2.53}$$

or

$$\mathbf{X} \cdot \mathbf{H} = [\mathbf{D} \quad \mathbf{C}]\mathbf{y_L}. \tag{2.54}$$

Here, the $N \times (2N-1)$ matrix in Eq. (2.53) has been partitioned into the $N \times N$ matrix $\mathbf{D}$, and the $N \times (N-1)$ matrix $\mathbf{C}$,

$$\mathbf{C} = \begin{bmatrix} z_0^{-N} & z_0^{-(N+1)} & \cdots & z_0^{-(2N-2)} \\ z_1^{-N} & z_1^{-(N+1)} & \cdots & z_1^{-(2N-2)} \\ \vdots & \vdots & \ddots & \vdots \\ z_{N-1}^{-N} & z_{N-1}^{-(N+1)} & \cdots & z_{N-1}^{-(2N-2)} \end{bmatrix}. \tag{2.55}$$

From Eqs. (2.50) and (2.54), we obtain

$$\mathbf{y_C} = \mathbf{D}^{-1}[\mathbf{D} \quad \mathbf{C}]\mathbf{y_L} = \left[\mathbf{I} \quad \mathbf{D}^{-1}\mathbf{C}\right]\mathbf{y_L}. \tag{2.56}$$

It is interesting to note that in the special case of the DFT, $\mathbf{D}$ is the $N \times N$ DFT matrix, and $\mathbf{C}$ comprises of the first $N-1$ columns of $\mathbf{D}$, due to the inherent periodicity. Thus,

$$\mathbf{D}^{-1}\mathbf{C} = \begin{bmatrix} 1 & 0 & 0 & \cdots & 0 & 0 \\ 0 & 1 & 0 & \cdots & 0 & 0 \\ \vdots & \vdots & \vdots & \ddots & \vdots & \vdots \\ 0 & 0 & 0 & \cdots & 0 & 1 \\ 0 & 0 & 0 & \cdots & 0 & 0 \end{bmatrix}. \tag{2.57}$$

In this case, we have the relation

$$\mathbf{y_C} = \begin{bmatrix} y[0] + y[N] \\ y[1] + y[N+1] \\ \vdots \\ y[N-2] + y[2N-2] \\ y[N-1] \end{bmatrix}, \tag{2.58}$$

which is simply a time-aliased version of the linear convolution $\mathbf{y_L}$, as expected. However, in the general case of the NDFT, the relationship between $\mathbf{y_C}$ and $\mathbf{y}$ is more complex, since the matrix $\mathbf{D}^{-1}\mathbf{C}$ does not have a simple structure such as Eq. (2.57).

Now, let us augment sequences $\mathbf{x}$ and $\mathbf{h}$ with $N-1$ zeros at the end to get sequences $\mathbf{x_a}$ and $\mathbf{y_a}$, which have lengths $M = 2N-1$. Then, if we take M-point NDFTs of $\mathbf{x}$ and $\mathbf{h}$, we obtain

$$\mathbf{X_a} = \mathbf{D_a}\mathbf{x_a}, \qquad \mathbf{H_a} = \mathbf{D_a}\mathbf{h_a}, \tag{2.59}$$

where $\mathbf{D_a}$ is an $M \times M$ NDFT matrix with sampling points $z_0, z_1, \ldots, z_{M-1}$. If we now multiply the NDFTs $\mathbf{X_a}$ and $\mathbf{H_a}$ (componentwise), and take the inverse NDFT of the result, we have the sequence

$$\mathbf{y_{Ca}} = \mathbf{D_a}^{-1}[\mathbf{X_a} \cdot \mathbf{H_a}]. \tag{2.60}$$

To see how this sequence is related to $\mathbf{y}$, we evaluate the z-transform of both sides of Eq. (2.44) at $z = z_0, z_1, \ldots, z_{M-1}$ to obtain

$$\mathbf{X_a} \cdot \mathbf{H_a} = \mathbf{D_a y}. \tag{2.61}$$

Therefore,

$$\mathbf{y_L} = \mathbf{D_a}^{-1}[\mathbf{X_a} \cdot \mathbf{H_a}]. \tag{2.62}$$

Eqs. (2.60) and (2.62) show that $\mathbf{y_{Ca}}$ equals the linear convolution $\mathbf{y_L}$. The linear convolution convolution of two sequences can, therefore, be computed by zero-padding them, multiplying their NDFTs and taking the inverse NDFT of the result. This is similar to the process of computing linear convolution using the DFT [Mitra, 1998]. The above results also hold when the sequences $x[n]$ and $y[n]$ have unequal lengths, say N_1 and N_2, respectively. In this case, the only difference is that they have to be padded with zeros to obtain sequences of length $M = N_1 + N_2 - 1$ before their NDFTs are computed.

Example 2.4

Consider two length-3 sequences,

$$\begin{aligned} \{x[n]\} &= \{1 \quad 2 \quad 4\}, \\ \{h[n]\} &= \{3 \quad -2 \quad 1\}. \end{aligned}$$

The linear convolution of these two sequences is given by

$$\{y_L[n]\} = \{3 \quad 4 \quad 9 \quad -6 \quad 4\}.$$

We next compute the 3-point NDFTs of $\{x[n]\}$ and $\{h[n]\}$ evaluated at the points, $z_0 = \infty$, $z_1 = 1$, $z_2 = -1$. The 3×3 NDFT matrix $\mathbf{D}$ is given by

$$\mathbf{D} = \begin{bmatrix} 1 & z_0^{-1} & z_0^{-2} \\ 1 & z_1^{-1} & z_1^{-2} \\ 1 & z_2^{-1} & z_2^{-2} \end{bmatrix} = \begin{bmatrix} 1 & 0 & 0 \\ 1 & 1 & 1 \\ 1 & -1 & 1 \end{bmatrix},$$

and its inverse is given by

$$\mathbf{D} = \begin{bmatrix} 1 & 0 & 0 \\ 0 & \frac{1}{2} & -\frac{1}{2} \\ -1 & \frac{1}{2} & \frac{1}{2} \end{bmatrix}.$$

The pertinent NDFT matrices are, therefore, given by

$$\mathbf{X} = \begin{bmatrix} X[0] \\ X[1] \\ X[2] \end{bmatrix} = \begin{bmatrix} 1 & 0 & 0 \\ 1 & 1 & 1 \\ 1 & -1 & 1 \end{bmatrix} \begin{bmatrix} 1 \\ 2 \\ 4 \end{bmatrix} = \begin{bmatrix} 1 \\ 7 \\ 3 \end{bmatrix},$$

$$\mathbf{H} = \begin{bmatrix} H[0] \\ H[1] \\ H[2] \end{bmatrix} = \begin{bmatrix} 1 & 0 & 0 \\ 1 & 1 & 1 \\ 1 & -1 & 1 \end{bmatrix} \begin{bmatrix} 3 \\ -2 \\ 1 \end{bmatrix} = \begin{bmatrix} 3 \\ 2 \\ 6 \end{bmatrix},$$

and hence,

$$\mathbf{X}\cdot\mathbf{H} = \begin{bmatrix} 3 \\ 14 \\ 18 \end{bmatrix}.$$

From Eq. (2.50), we thus arrive at

$$\mathbf{y_C} = \begin{bmatrix} y_C[0] \\ y_C[1] \\ y_C[2] \end{bmatrix} = \mathbf{D^{-1}}[\mathbf{X}\cdot\mathbf{H}] = \begin{bmatrix} 1 & 0 & 0 \\ 0 & \frac{1}{2} & -\frac{1}{2} \\ -1 & \frac{1}{2} & \frac{1}{2} \end{bmatrix} \begin{bmatrix} 3 \\ 14 \\ 18 \end{bmatrix} = \begin{bmatrix} 3 \\ -2 \\ 13 \end{bmatrix}.$$

Now, from Eq. (2.53), we have

$$\begin{aligned} \mathbf{X}\cdot\mathbf{H} &= \begin{bmatrix} 1 & z_0^{-1} & z_0^{-2} & z_0^{-3} & z_0^{-4} \\ 1 & z_1^{-1} & z_1^{-2} & z_1^{-3} & z_1^{-4} \\ 1 & z_2^{-1} & z_2^{-2} & z_2^{-3} & z_2^{-4} \end{bmatrix} \begin{bmatrix} y_L[0] \\ y_L[1] \\ y_L[2] \\ y_L[3] \\ y_L[4] \end{bmatrix} \\ &= \begin{bmatrix} 1 & 0 & 0 & 0 & 0 \\ 1 & 1 & 1 & 1 & 1 \\ 1 & -1 & 1 & -1 & 1 \end{bmatrix} \begin{bmatrix} 3 \\ 4 \\ 9 \\ -6 \\ 4 \end{bmatrix} = \begin{bmatrix} 3 \\ -2 \\ 13 \end{bmatrix}, \end{aligned}$$

which is identical to the vector $\mathbf{y_C}$ derived earlier using the NDFT approach.

In order to determine the linear convolution of $\{x[n]\}$ and $\{h[n]\}$ using the NDFT, we augment both sequences with $N-1 = 2$ zeros, leading to the sequences,

$$\begin{aligned} \{x_a[n]\} &= \{1 \;\; 2 \;\; 4 \;\; 0 \;\; 0\}, \\ \{h_a[n]\} &= \{3 \;\; -2 \;\; 1 \;\; 0 \;\; 0\}. \end{aligned}$$

We then compute the 5-point NDFT of the two augmented sequences at the points, $z_0 = \infty$, $z_1 = 1$, and $z_2 = -1$, $z_3 = \frac{1}{2}$, and $z_4 = -\frac{1}{2}$. The 5×5 NDFT matrix is given by

$$\mathbf{D_a} = \begin{bmatrix} 1 & z_0^{-1} & z_0^{-2} & z_0^{-3} & z_0^{-4} \\ 1 & z_1^{-1} & z_1^{-2} & z_1^{-3} & z_1^{-4} \\ 1 & z_2^{-1} & z_2^{-2} & z_2^{-3} & z_2^{-4} \\ 1 & z_3^{-1} & z_3^{-2} & z_3^{-3} & z_3^{-4} \\ 1 & z_4^{-1} & z_4^{-2} & z_4^{-3} & z_4^{-4} \end{bmatrix} = \begin{bmatrix} 1 & 0 & 0 & 0 & 0 \\ 1 & 1 & 1 & 1 & 1 \\ 1 & -1 & 1 & -1 & 1 \\ 1 & 2 & 4 & 8 & 16 \\ 1 & -2 & 4 & -8 & 16 \end{bmatrix},$$

and the inverse NDFT is given by

$$\mathbf{D_a}^{-1} = \begin{bmatrix} 1 & 0 & 0 & 0 & 0 \\ 0 & \frac{2}{3} & -\frac{2}{3} & -\frac{1}{12} & \frac{1}{12} \\ -\frac{5}{4} & \frac{2}{3} & \frac{2}{3} & -\frac{1}{24} & -\frac{1}{24} \\ 0 & -\frac{1}{6} & \frac{1}{6} & \frac{1}{12} & -\frac{1}{12} \\ \frac{1}{4} & -\frac{1}{6} & -\frac{1}{6} & \frac{1}{24} & \frac{1}{24} \end{bmatrix}.$$

The corresponding NDFT matrices for $\{x_a[n]\}$ and $\{h_a[n]\}$ are

$$\mathbf{X_a} = \begin{bmatrix} X[0] \\ X[1] \\ X[2] \\ X[3] \\ X[4] \end{bmatrix} = \begin{bmatrix} 1 & 0 & 0 & 0 & 0 \\ 1 & 1 & 1 & 1 & 1 \\ 1 & -1 & 1 & -1 & 1 \\ 1 & 2 & 4 & 8 & 16 \\ 1 & -2 & 4 & -8 & 16 \end{bmatrix} \begin{bmatrix} 1 \\ 2 \\ 4 \\ 0 \\ 0 \end{bmatrix} = \begin{bmatrix} 1 \\ 7 \\ 3 \\ 21 \\ 13 \end{bmatrix},$$

$$\mathbf{H_a} = \begin{bmatrix} H[0] \\ H[1] \\ H[2] \\ H[3] \\ H[4] \end{bmatrix} = \begin{bmatrix} 1 & 0 & 0 & 0 & 0 \\ 1 & 1 & 1 & 1 & 1 \\ 1 & -1 & 1 & -1 & 1 \\ 1 & 2 & 4 & 8 & 16 \\ 1 & -2 & 4 & -8 & 16 \end{bmatrix} \begin{bmatrix} 3 \\ -2 \\ 1 \\ 0 \\ 0 \end{bmatrix} = \begin{bmatrix} 3 \\ 2 \\ 6 \\ 3 \\ 11 \end{bmatrix},$$

and hence,

$$\mathbf{X_a} \cdot \mathbf{H_a} = \begin{bmatrix} 3 \\ 14 \\ 18 \\ 63 \\ 143 \end{bmatrix}.$$

From Eq. (2.62), the linear convolution of $\{x[n]\}$ and $\{h[n]\}$ is given by

$$\begin{aligned} \mathbf{y_L} &= \mathbf{D_a}^{-1}[\mathbf{X_a} \cdot \mathbf{H_a}] \\ &= \begin{bmatrix} 1 & 0 & 0 & 0 & 0 \\ 0 & \frac{2}{3} & -\frac{2}{3} & -\frac{1}{12} & \frac{1}{12} \\ -\frac{5}{4} & \frac{2}{3} & \frac{2}{3} & -\frac{1}{24} & -\frac{1}{24} \\ 0 & -\frac{1}{6} & \frac{1}{6} & \frac{1}{12} & -\frac{1}{12} \\ \frac{1}{4} & -\frac{1}{6} & -\frac{1}{6} & \frac{1}{24} & \frac{1}{24} \end{bmatrix} \begin{bmatrix} 3 \\ 14 \\ 18 \\ 63 \\ 143 \end{bmatrix} = \begin{bmatrix} 3 \\ 4 \\ 9 \\ -6 \\ 4 \end{bmatrix}, \end{aligned}$$

as expected.

2.3 COMPUTATION OF THE NDFT

We address the problem of computing the NDFT in this section. The factors to be considered are—the *amount of computation* needed in terms of multiplications and additions, and the *number of coefficients* used for computation. To establish a reference for comparing these factors, we begin by examining the direct method. This is followed by Horner's method, which requires the same amount of computation as the direct method, but a lower number of coefficients. Finally, we use the Goertzel algorithm [Goertzel, 1958] to compute the NDFT. We interpret this algorithm in two ways—computing the output of a digital filter, and evaluating a finite trigonometric series. The Goertzel algorithm achieves a reduction in computation as well as the number of coefficients used. The computational saving is obtained when the NDFT is evaluated at points on the unit circle in the z-plane.

Due to the *generalized sampling* inherent in the NDFT, there is no periodicity or symmetry in the complex numbers, z_k^{-n}. Consequently, algorithms as efficient as the FFT cannot be derived in the general case. In the next section, we will present the subband NDFT, which leads to a fast, approximate computation of the NDFT, when the signal has its energy concentrated in a few bands of the spectrum.

2.3.1 Direct Method

We can compute the NDFT directly by evaluating the expression given in Eq. (2.1). In general, the signal $x[n]$ is a complex sequence of length N, and the sampling points $z_0, z_1, \ldots, z_{N-1}$ are also complex numbers. To compute each sample of the NDFT, we need N complex multiplications and $(N-1)$ complex additions, i.e., $4N$ real multiplications and $(4N-2)$ real additions. Therefore, the amount of computation needed for evaluating N samples of the NDFT is approximately proportional to N^2. Note that this is the same as that needed for directly computing the DFT. In the direct method, we need to compute or store N complex coefficients, $\{z_k^{-n}, \quad n = 0, 1, \ldots, N-1\}$ for evaluating each NDFT sample $X(z_k)$.

2.3.2 Horner's Method

Let us rewrite the NDFT in Eq. (2.1) as

$$X(z_k) = z_k^{-(N-1)} A_k, \tag{2.63}$$

where

$$\begin{aligned} A_k &= \sum_{n=0}^{N-1} x[n] z_k^{N-1-n} \\ &= x[0] z_k^{N-1} + x[1] z_k^{N-2} + \cdots + x[N-2] z_k + x[N-1]. \end{aligned} \tag{2.64}$$

To avoid using the N coefficients z_k^n, we can express Eq. (2.64) as a nested multiplication (also known as Horner's method [Atkinson, 1978]):

$$A_k = \{\cdots (x[0] z_k + x[1]) z_k + \cdots\} z_k + x[N-1]. \tag{2.65}$$

We start by evaluating the expression in the innermost parentheses of Eq. (2.65), and proceed to solve for A_k. $X(z_k)$ can then be found from

Eq. (2.63). This requires a total of $4N$ real multiplications and $(4N - 2)$ real additions, which is the same as in the direct method. However, in Horner's method, we need *only two coefficients*, z_k and $z_k^{-(N-1)}$, to evaluate the NDFT sample $X(z_k)$.

Note that the NDFT has been expressed as shown in Eq. (2.63), so that we can start evaluating the nested multiplication in Eq. (2.65) with the first sample $x[0]$, rather than the last sample $x[N-1]$. This eliminates the need for buffering the input signal.

2.3.3 Goertzel Algorithm: Digital Filter Interpretation

Consider the NDFT sample at $z = z_k$, given by

$$X(z_k) = \sum_{r=0}^{N-1} x[r] z_k^{-r}. \tag{2.66}$$

Multiplying both sides of Eq. (2.66) by z_k^N, we have

$$z_k^N X(z_k) = \sum_{r=0}^{N-1} x[r] z_k^{N-r}. \tag{2.67}$$

Let us now define the sequence

$$y_k[n] = \sum_{r=-\infty}^{\infty} x[r] z_k^{n-r} u[n-r], \tag{2.68}$$

where $u[n]$ denotes the unit step sequence. Eq. (2.68) is equivalent to the *discrete convolution*

$$y_k[n] = x[n] * z_k^n u[n]. \tag{2.69}$$

Since $x[n]$ is zero outside the range $0 \leq n \leq N-1$, we infer from Eqs. (2.67) and (2.68) that

$$X(z_k) = z_k^{-N} \, y_k[n]|_{n=N}. \tag{2.70}$$

Thus, the NDFT sample $X(z_k)$ is obtained by multiplying z_k^{-N} with the Nth sample at the output of a system, whose impulse response is $z_k^n u[n]$. This is a first-order recursive system, whose system function is

$$H_k(z) = \frac{1}{1 - z_k z^{-1}}. \tag{2.71}$$

To compute $X(z_k)$ using this system, we need $(4N+4)$ real multiplications and $(4N+2)$ real additions. These are nearly the same as in the direct method. However, only *two coefficients*, z_k and z_k^{-N}, are needed.

Next, we consider a special case for which the number of multiplications can be *reduced* by a factor of two. Multiplying the numerator and denominator of Eq. (2.71) by $(1 - z_k^* z^{-1})$, where z_k^* is the complex conjugate of z_k, we obtain

$$H_k(z) = \frac{1 - z_k^* z^{-1}}{1 - 2\,Re\{z_k\} z^{-1} + |z_k|^2 z^{-2}}. \tag{2.72}$$

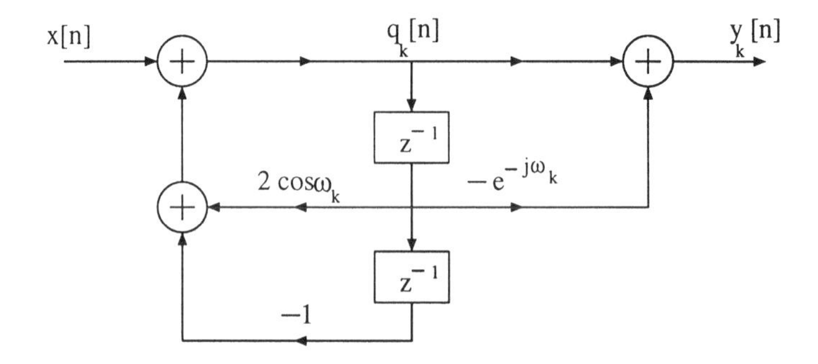

Figure 2.2. Goertzel algorithm as a second-order recursive computation.

Let $|z_k|^2 = 1$. Thus, the NDFT is being evaluated at a point $z_k = e^{j\omega_k}$ on the unit circle in the z-plane. In this case, we can simplify Eq. (2.72) to obtain the system function

$$H_k(z) = \frac{1 - e^{-j\omega_k} z^{-1}}{1 - 2\cos\omega_k \, z^{-1} + z^{-2}}, \tag{2.73}$$

which is shown in Fig. 2.2. Thus, we can interpret the Goertzel algorithm as computing the output of this second-order recursive digital filter.

Since we want to compute only $y_k[N]$, the multiplication by $e^{-j\omega_k}$ in the feedforward section of Fig. 2.2 (to the right of the delay elements) need not be performed until the Nth iteration. The intermediate signals, $q_k[n]$ and $q_k[n-1]$, are computed recursively using the difference equation

$$q_k[n] = a_k \, q_k[n-1] - q_k[n-2] + x[n], \qquad n = 0, 1, \ldots, N, \tag{2.74}$$

where the coefficient a_k is given by

$$a_k = 2\cos\omega_k,$$

and the initial conditions are

$$q_k[-1] = q_k[-2] = 0.$$

Finally, we evaluate $y_k[N]$ as follows:

$$y_k[N] = q_k[N] - e^{-j\omega_k} \, q_k[N-1]. \tag{2.75}$$

The total computation required for $X(z_k)$ is $(2N + 8)$ real multiplications and $(4N + 6)$ real additions. Thus, the number of multiplications is nearly *half* of that in the direct method. In addition, we need only two coefficients, $e^{-j\omega_k}$ and $e^{-j\omega_k N}$.

As expected, the Goertzel algorithm described here reduces to the one used for computing the DFT, when $\omega_k = 2\pi k/N$. The amount of computation required for the DFT and NDFT is nearly the same. Only one extra complex multiplication is needed in the case of the NDFT.

In some applications, we are interested in finding only the *squared magnitude* of the NDFT, i.e., $|X(z_k)|^2$. In this case, the Goertzel algorithm can be *modified* as follows. Since $z_k = e^{j\omega_k}$, we infer from Eq. (2.70) that

$$|X(z_k)|^2 = |y_k[N]|^2. \tag{2.76}$$

Substituting for $y_k[N]$ from Eq. (2.75), we obtain

$$|y_k[N]|^2 = q_k^2[N] + q_k^2[N-1] + a_k\, q_k[N]\, q_k[N-1]. \tag{2.77}$$

This modified scheme uses only one real coefficient a_k. If we have a real input signal, then complex arithmetic is avoided, and we need only $(N+4)$ real multiplications and $(2N+2)$ real additions. Note that this is equal to the computation needed for finding the squared magnitude of the DFT. This scheme will be used in Chapter 6 to detect dual-tone multi-frequency (DTMF) tones.

2.3.4 *Goertzel Algorithm: Trigonometric Series Interpretation*

The algorithm originally proposed by Goertzel [Goertzel, 1958] was a method for evaluation of finite trigonometric series. We have arrived at this interpretation independently, while attempting to compute the NDFT at points on the unit circle in the z-plane.

We can express the NDFT sample at $z_k = e^{j\omega_k}$ as

$$X(z_k) = \sum_{n=0}^{N-1} x[n] e^{-j\omega_k n} = e^{-j(N-1)\omega_k} A_k, \tag{2.78}$$

where

$$A_k = \sum_{n=0}^{N-1} x[n] e^{j(N-1-n)\omega_k}. \tag{2.79}$$

From Eq. (2.79), we can decompose A_k as

$$A_k = C_k + jS_k, \tag{2.80}$$

where

$$C_k = \sum_{n=0}^{N-1} x[n]\cos(N-1-n)\omega_k, \tag{2.81}$$

and

$$S_k = \sum_{n=0}^{N-1} x[n]\sin(N-1-n)\omega_k. \tag{2.82}$$

The direct computation of C_k and S_k from Eqs. (2.81) and (2.82) requires the values of the $2N$ coefficients $\cos n\omega_k$ and $\sin n\omega_k$. To avoid this, we attempt to find an iterative computation method which uses only the values of $\cos\omega_k$ and $\sin\omega_k$. To this end, we define the partial sums

$$\begin{aligned} C_k^{(i)} &= \sum_{n=0}^{i} x[n]\cos(N-1-n)\omega_k, \\ S_k^{(i)} &= \sum_{n=0}^{i} x[n]\sin(N-1-n)\omega_k, \quad i=0,1,\ldots,N-1, \end{aligned} \tag{2.83}$$

so that

$$C_k = C_k^{(N-1)}, \quad S_k = S_k^{(N-1)}. \tag{2.84}$$

Focusing our attention on C_k first, we observe that

$$C_k^{(1)} = x[0]\cos(N-1)\omega_k + x[1]\cos(N-2)\omega_k. \tag{2.85}$$

By using the trigonometric relation

$$\cos(m+1)\theta = 2\cos\theta\,\cos m\theta - \cos(m-1)\theta, \tag{2.86}$$

we can express Eq. (2.85) as

$$C_k^{(1)} = \{2\cos\omega_k\, x[0] + x[1]\}\cos(N-2)\omega_k - x[0]\cos(N-3)\omega_k. \tag{2.87}$$

The next partial sum can then be expressed as

$$\begin{aligned} C_k^{(2)} &= C_k^{(1)} + x[2]\cos(N-3)\omega_k \\ &= \{2\cos\omega_k\, x[0] + x[1]\}\cos(N-2)\omega_k \\ &\quad +\{x[2] - x[0]\}\cos(N-3)\omega_k. \end{aligned} \tag{2.88}$$

We can continue doing this, i.e., substituting for the cosine of the mth harmonic with those of the two lower harmonics, and forming the successive partial sums, to obtain

$$C_k^{(i)} = g_i\cos(N-i)\omega_k + h_i\cos(N-i-1)\omega_k, \quad i=1,2,\ldots,N-1, \tag{2.89}$$

where

$$\begin{aligned} g_1 &= x[0], \quad h_1 = x[1], \\ g_i &= 2\cos\omega_k\, g_{i-1} + h_{i-1}, \quad h_i = x[i] - g_{i-1}, \quad i=2,3,\ldots,N-1. \end{aligned} \tag{2.90}$$

Therefore, we can compute C_k by recursively solving for g_i and h_i, since

$$C_k = C_k^{(N-1)} = g_{N-1}\cos\omega_k + h_{N-1}. \tag{2.91}$$

Similarly, to find S_k, we use the relation

$$\sin(m+1)\theta = 2\cos\theta\,\sin m\theta - \sin(m-1)\theta, \tag{2.92}$$

to express the partial sums $S_k^{(i)}$ as

$$S_k^{(i)} = g_i \sin(N-i)\omega_k + h_i \sin(N-i-1)\omega_k, \quad i = 1, 2, \ldots, N-1. \tag{2.93}$$

This implies that

$$S_k = S_k^{(N-1)} = g_{N-1} \sin \omega_k. \tag{2.94}$$

Therefore, we can also compute S_k from the values already found for g_i and h_i. Finally, we compute A_k using Eq. (2.80), and therefore, $X(z_k)$ from Eq. (2.78). The total computation is $(2N+4)$ real multiplications and $(4N-2)$ real additions. Besides, we need only the coefficients, $\cos\omega_k$, $\sin\omega_k$, and $e^{-j(N-1)\omega_k}$. If the input signal is real, C_k and S_k are also real. Therefore, the total computation reduces to $(N+2)$ real multiplications and $(2N-1)$ real additions.

2.4 SUBBAND NDFT

In this section, we outline the subband NDFT (SB-NDFT), which is a method for computing the NDFT based on a subband decomposition of the input sequence. This is a generalization of the subband DFT (SB-DFT) which was discussed in Section 1.4 [Shentov et al., 1995]. The method is useful for developing *fast algorithms* to approximately compute NDFT samples for signals which have their energy concentrated in only a few bands of the spectrum. We shall use such an algorithm based on the SB-NDFT in Chapter 6, for decoding DTMF tones.

We begin by presenting the basic idea of the SB-NDFT for a full-band analysis. Then, we show how to achieve reduced computation by partial-band analysis. Finally, we incorporate a generalized pre-processing stage in the SB-NDFT.

2.4.1 Full-Band Analysis

Consider a signal $x[n]$ with an even number of samples, N. We first decompose $x[n]$ into two subsequences, $g_L[n]$ and $g_H[n]$, of length $N/2$ each, as given by Eq. (1.17). These sequences are simply lowpass and highpass filtered, and downsampled versions of $x[n]$. We can express $x[n]$ in terms of $g_L[n]$ and $g_H[n]$ as shown in Eq. (1.18). Then, the z-transform $X(z)$ of $x[n]$ can be expressed as in Eq. (1.19), in terms of $G_L(z)$ and $G_H(z)$, the z-transforms of $g_L[n]$ and $g_H[n]$.

By evaluating Eq. (1.19) at $z = z_k, k = 0, 1, \ldots, N-1$, we obtain the NDFT $X(z_k)$ of $x[n]$,

$$X(z_k) = (1 + z_k^{-1})\, G_L(z_k^2) + (1 - z_k^{-1})\, G_H(z_k^2), \tag{2.95}$$

where $G_L(z_k^2)$ and $G_H(z_k^2)$ are the NDFTs of the subsequences, $g_L[n]$ and $g_H[n]$, evaluated at the points $z = z_k^2$. Therefore, we can find the NDFT of $x[n]$ at the points $z = z_k$ by computing the NDFTs of the smaller sequences, $g_L[n]$ and $g_H[n]$, at the points $z = z_k^2$, and then combining them as shown in Eq. (2.95). Fig. 2.3 shows a SB-NDFT computation with this two-band decomposition.

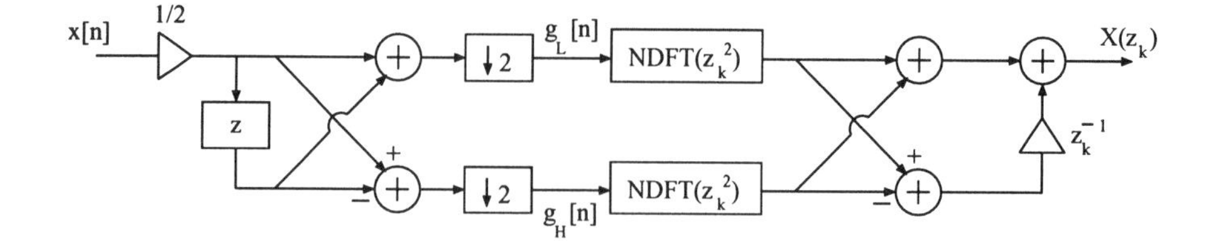

Figure 2.3. SB-NDFT computation with a two-band decomposition.

Eq. (2.95) can be expressed in a matrix form as

$$X(z_k) = [1 \quad z_k^{-1}]\, \mathbf{R_2} \begin{bmatrix} G_L(z_k^2) \\ G_H(z_k^2) \end{bmatrix}, \tag{2.96}$$

where $\mathbf{R_2}$ is the 2×2 Hadamard matrix

$$\mathbf{R_2} = \begin{bmatrix} 1 & 1 \\ 1 & -1 \end{bmatrix}. \tag{2.97}$$

We can repeat the decomposition in Eq. (2.95) for each of the two subsequences, and continue similarly as long as the lengths of the subsequences are even. For example, a two-stage decomposition yields

$$\begin{aligned} X(z_k) &= (1 + z_k^{-1})\{(1 + z_k^{-2})\, G_{LL}(z_k^4) + (1 - z_k^{-2})\, G_{HL}(z_k^4)\} \\ &\quad +(1 - z_k^{-1})\{(1 + z_k^{-2})\, G_{LH}(z_k^4) + (1 - z_k^{-2}) G_{HH}\,(z_k^4)\} \\ &= [1 \quad z_k^{-1} \quad z_k^{-2} \quad z_k^{-3}]\, \mathbf{R_4} \begin{bmatrix} G_{LL}(z_k^4) \\ G_{LH}(z_k^4) \\ G_{HL}(z_k^4) \\ G_{HH}(z_k^4) \end{bmatrix}, \end{aligned} \tag{2.98}$$

where $\mathbf{R_4}$ is the 4×4 Hadamard matrix

$$\mathbf{R_4} = \begin{bmatrix} 1 & 1 & 1 & 1 \\ 1 & -1 & 1 & -1 \\ 1 & 1 & -1 & -1 \\ 1 & -1 & -1 & 1 \end{bmatrix} = \mathbf{R_2} \otimes \mathbf{R_2}, \tag{2.99}$$

with $\otimes$ denoting the Kronecker product [Regalia and Mitra, 1989]. If the number of samples in the input signal is $N = MP$, where $M = 2^\mu$, we can perform a μ-stage decomposition. The filtering and downsampling operations can be reorganized into a preprocessing stage as shown in Fig. 2.4. Here, all the filtering and downsampling steps have been combined by taking a block

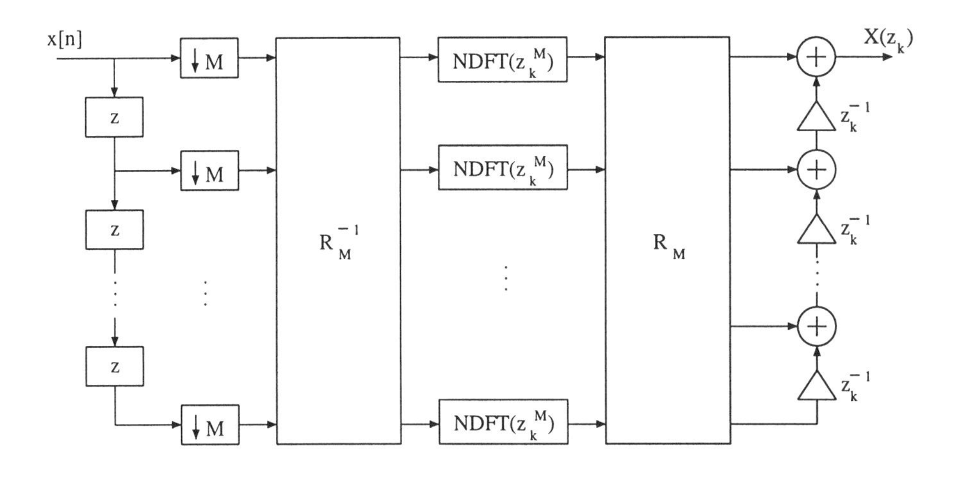

Figure 2.4. SB-NDFT computation with an M-band decomposition.

transform, every M samples. This transform can be recognized as the $M \times M$ inverse Hadamard matrix

$$\mathbf{R}_\mathbf{M}^{-1} = \frac{1}{M}\mathbf{R}_\mathbf{M}, \tag{2.100}$$

where

$$\mathbf{R}_\mathbf{M} = \mathbf{R}_{\mathbf{M/2}} \otimes \mathbf{R}_\mathbf{2}. \tag{2.101}$$

The NDFTs of the subsequences can be implemented by any of the methods discussed in Section 2.3. These NDFTs are fed to a recombination stage, composed of an $M \times M$ Hadamard transform and a chain of multipliers and adders.

In the full-band case, the SB-NDFT realization requires slightly more computation as compared with the usual NDFT. For example, in the case of the two-band decomposition, the evaluation of the NDFT sample of a sequence of length N at $z = z_k$ requires finding NDFT samples of two $N/2$-length sequences at $z = z_k^2$. For this part of the calculation, we need the same amount of computation as the usual NDFT. But extra computation is required to implement the preprocessing stage (additions) and the recombination stage (multiplications and additions). However, the spectral separation achieved in the SB-NDFT enables us to obtain *substantial* computational savings when the signal has energy concentrated in only a few bands of the spectrum. This is discussed next in a partial-band analysis.

2.4.2 Partial-Band Analysis

As in the case of the SB-DFT [Shentov et al., 1991], we can compute the NDFT approximately in the primary band of a signal, by utilizing the spectral separation provided by the preprocessing stage. This is done by discarding calculations in bands which have a negligible fraction of the total energy. For example, if a signal has most of its energy in the low frequency band $0 \leq f \leq f_s/4$, we can obtain a reasonable approximation to the NDFT in Eq. (2.95) by dropping the highpass term $G_H(z_k^2)$, as follows:

$$\hat{X}(z_k) = (1 + z_k^{-1})\, G_L(z_k^2), \tag{2.102}$$

where $z_k = e^{j\omega_k}$. Note that the amount of computation required here is nearly *half* of that in Eq. (2.95). In a general M-band decomposition, if L of the M bands are discarded, the computational saving is approximately L/M.

The decision for discarding low-energy bands can be taken in real-time by performing primary band estimation using the technique discussed in [Shentov and Mitra, 1991]. Note that the approximate NDFT samples suffer from two types of error—*linear distortion* due to non-constant frequency responses in the band of interest, and *aliasing* due to frequency components in the discarded band. As an example, we examine the error in the two-band case, when the NDFT is approximated by Eq. (2.102). In an effort to compensate for linear distortion, we rewrite this equation as

$$\hat{X}(z_k) = \frac{1}{2}(1 + z_k^{-1}) \sum_{n=0}^{N/2-1} \{x[2n] + x[2n+1]\} z_k^{-2n}. \tag{2.103}$$

We can rearrange the terms to express this as the sum of the actual NDFT $X(z_k)$ and an aliasing component $X_1(z_k)$,

$$\hat{X}(z_k) = X(z_k) + X_1(z_k). \tag{2.104}$$

Since decimation causes repeated spectra along the frequency axis, $\hat{X}(z_k)$ can also be expressed as

$$\hat{X}(z_k) = A_0 X(z_k) + A_1 X(-z_k), \tag{2.105}$$

where A_0 and A_1 are coefficients possibly depending on z_k. The component $X(-z_k)$ arises because of a two-fold decimation. Using Eq. (2.1) to substitute for $X(z_k)$ and $X(-z_k)$, we have

$$\begin{aligned} \hat{X}(z_k) &= \sum_{n=0}^{N-1} \{A_0 + (-1)^n A_1\}\, x[n] z_k^{-n} \\ &= \sum_{n=0}^{N/2-1} (A_0 + A_1)\, x[2n] z_k^{-2n} \\ &\quad + z_k^{-1} \sum_{n=0}^{N/2-1} (A_0 - A_1)\, x[2n+1] z_k^{-2n}. \end{aligned} \tag{2.106}$$

By comparing Eqs. (2.103) and (2.106), we infer that

$$\begin{aligned} A_0 + A_1 &= \frac{1}{2}(1 + z_k^{-1}), \\ A_0 - A_1 &= \frac{1}{2}(1 + z_k). \end{aligned} \tag{2.107}$$

Thus, we can solve for A_0 and A_1. Using these results in Eq. (2.105), with $z_k = e^{j\omega_k}$, we obtain

$$\hat{X}(e^{j\omega_k}) = \frac{1}{2}(1 + \cos\omega_k)\, X(e^{j\omega_k}) - \frac{j}{2}\sin\omega_k\, X(-e^{j\omega_k}). \tag{2.108}$$

Therefore, the approximation error is

$$E(z_k) = X(z_k) - \hat{X}(z_k) = \frac{1}{2}(1 - \cos\omega_k)\, X(z_k) + \frac{j}{2}\sin\omega_k\, X(-z_k). \tag{2.109}$$

This procedure can also be used to find the error for a larger order of decomposition and a different number of retained bands.

If the signal has no energy beyond $0 \le f \le f_s/4$, the second term in Eq. (2.108) can be dropped. The exact NDFT can then be obtained by scaling:

$$X(z_k) = \frac{2\hat{X}(z_k)}{1 + \cos\omega_k}. \tag{2.110}$$

Thus, we can *fully compensate* for all linear distortion in the band of interest.

2.4.3 *Generalized Preprocessing*

If the number of samples in the input signal is $N = MP$, where $M = 2^\mu$, we can obtain a μ-stage SB-NDFT as

$$X(z_k) = \begin{bmatrix} 1 & z_k^{-1} & \cdots & z_k^{-(M-1)} \end{bmatrix} \mathbf{R_M} \begin{bmatrix} G_0(z_k^M) \\ G_1(z_k^M) \\ \vdots \\ G_{M-1}(z_k^M) \end{bmatrix}, \tag{2.111}$$

where $\mathbf{R_M}$ is the $M \times M$ Hadamard matrix, and $G_i(z_k^M)$ is the NDFT of the subsequence $g_i[n]$ at the point $z = z_k^M$. From Fig. 2.4, it is clear that the subsequences are related to the polyphase components of $x[n]$ as follows:

$$\begin{bmatrix} g_0[n] \\ g_1[n] \\ \vdots \\ g_{M-1}[n] \end{bmatrix} = \mathbf{R_M}^{-1} \begin{bmatrix} x[Mn] \\ x[Mn+1] \\ \vdots \\ x[Mn+M-1] \end{bmatrix}, \qquad n = 0, 1, \ldots, P-1. \tag{2.112}$$

The SB-NDFT can be *generalized* by replacing $\mathbf{R_M}$ with an $M \times M$ nonsingular matrix $\mathbf{C}$. Doing this, we obtain

$$X(z_k) = \begin{bmatrix} 1 & z_k^{-1} & \cdots & z_k^{-(M-1)} \end{bmatrix} \mathbf{C} \begin{bmatrix} V_0(z_k^M) \\ V_1(z_k^M) \\ \vdots \\ V_{M-1}(z_k^M) \end{bmatrix}, \tag{2.113}$$

where $V_i(z_k^M)$ is the NDFT of the subsequence $v_i[n]$ at $z = z_k^M$. The subsequences $v_i[n]$ are given by

$$\begin{bmatrix} v_0[n] \\ v_1[n] \\ \vdots \\ v_{M-1}[n] \end{bmatrix} = \mathbf{C}^{-1} \begin{bmatrix} x[Mn] \\ x[Mn+1] \\ \vdots \\ x[Mn+M-1] \end{bmatrix}, \qquad n = 0, 1, \ldots, P-1. \tag{2.114}$$

In the general case, M need not be a power of two. The poor frequency separation given by the Hadamard transform can be improved by using other matrices. Simple multiplierless transforms are usually preferred since they offer low computational complexity.

2.5 THE 2-D NDFT

2.5.1 Definition

As in the case of a 1-D sequence, the *nonuniform discrete Fourier transform* of a 2-D sequence corresponds to sampling its 2-D z-transform. The 2-D NDFT of a sequence $x[n_1, n_2]$ of size $N_1 \times N_2$ is defined as

$$\hat{X}(z_{1k}, z_{2k}) = \sum_{n_1=0}^{N_1-1} \sum_{n_2=0}^{N_2-1} x[n_1, n_2] z_{1k}^{-n_1} z_{2k}^{-n_2}, \qquad k = 0, 1, \ldots, N_1 N_2 - 1, \tag{2.115}$$

where (z_{1k}, z_{2k}) represent $N_1 N_2$ distinct points in the 4-D (z_1, z_2) space. These points can be chosen arbitrarily but in such a way that the inverse transform exists.

We illustrate this by a simple example. Consider the case, $N_1 = N_2 = 2$. We can express Eq. (2.115) in a matrix form as

$$\hat{\mathbf{X}} = \mathbf{D}\mathbf{X}, \tag{2.116}$$

where

$$\hat{\mathbf{X}} = \begin{bmatrix} \hat{X}(z_{10}, z_{20}) \\ \hat{X}(z_{11}, z_{21}) \\ \hat{X}(z_{12}, z_{22}) \\ \hat{X}(z_{13}, z_{23}) \end{bmatrix}, \qquad \mathbf{X} = \begin{bmatrix} x[0,0] \\ x[0,1] \\ x[1,0] \\ x[1,1] \end{bmatrix}, \tag{2.117}$$

and

$$\mathbf{D} = \begin{bmatrix} 1 & z_{20}^{-1} & z_{10}^{-1} & z_{10}^{-1} z_{20}^{-1} \\ 1 & z_{21}^{-1} & z_{11}^{-1} & z_{11}^{-1} z_{21}^{-1} \\ 1 & z_{22}^{-1} & z_{12}^{-1} & z_{12}^{-1} z_{22}^{-1} \\ 1 & z_{23}^{-1} & z_{13}^{-1} & z_{13}^{-1} z_{23}^{-1} \end{bmatrix}. \tag{2.118}$$

In general, the 2-D NDFT matrix $\mathbf{D}$ is of size $N_1 N_2 \times N_1 N_2$. It is fully specified by the choice of the $N_1 N_2$ sampling points. For the 2-D NDFT matrix to exist uniquely, these points should be chosen so that $\mathbf{D}$ is nonsingular. In the case of the 1-D NDFT, if the points z_k are distinct, the inverse NDFT is guaranteed to exist uniquely. This happens because the 1-D NDFT matrix has a determinant that can always be factored as in Eq. (2.5). However, there is no simple extension of this to the 2-D case. Even if the 2-D sampling points are distinct, this does not guarantee the nonsingularity of $\mathbf{D}$. This calls for the need to make a judicious choice of sampling points. Some results have been derived on sufficient conditions under which the equivalent 2-D polynomial interpolation problem has a unique or nonunique solution when the samples are located on irreducible curves [Zakhor and Alvstad, 1992]. However, no set of necessary and sufficient conditions has been found. This does not pose a serious problem from the point of view of applications. For all practical purposes, we can just perform a check on the determinant of $\mathbf{D}$ to ascertain that it is nonzero for that particular choice of points. In the general case, the inverse 2-D NDFT is computed by solving a linear system of size N_1N_2, which requires $O(N_1^3N_2^3)$ operations.

Note that the definition of the NDFT can be readily extended to higher dimensions.

2.5.2 Special Cases of 2-D NDFT

In general, the determinant of the 2-D NDFT matrix is not factorizable. However, we now consider special cases in which the determinant can be factored. In these cases, the choice of the sampling points is restricted in some way so that the 2-D NDFT matrix is guaranteed to be nonsingular.

(1) Nonuniformly spaced rectangular grid

In this case, the sampling points lie at the vertices of a rectangular grid in the (z_1, z_2) space. For a sequence of size $N_1 \times N_2$, the z_1-coordinates of the N_1 grid lines running parallel to the z_2 axis can be chosen arbitrarily, so long as they are distinct. Let these coordinates be denoted by $z_{10}, z_{11}, \ldots, z_{1,N_1-1}$. Similarly, the z_2-coordinates of the N_2 grid lines running parallel to the z_1 axis can be chosen arbitrarily, so long as they are distinct. Let these coordinates be denoted by $z_{20}, z_{21}, \ldots, z_{2,N_2-1}$. This distribution of points is illustrated in Fig. 2.5, for an example where $N_1 = 3$, $N_2 = 4$. Note that the representation of the z_1 and z_2 axes in this figure is for convenience only, since the complex variables z_1 and z_2 actually form a 4-D space.

Eq. (2.116) can then be expressed in a simpler matrix form as

$$\hat{\mathbf{X}} = \mathbf{D}_1\mathbf{X}\mathbf{D}_2^t, \tag{2.119}$$

where

$$\hat{\mathbf{X}} = \begin{bmatrix} \hat{X}(z_{10}, z_{20}) & \hat{X}(z_{10}, z_{21}) & \cdots & \hat{X}(z_{10}, z_{2,N_2-1}) \\ \hat{X}(z_{11}, z_{20}) & \hat{X}(z_{11}, z_{21}) & \cdots & \hat{X}(z_{11}, z_{2,N_2-1}) \\ \vdots & \vdots & \ddots & \vdots \\ \hat{X}(z_{1,N_1-1}, z_{20}) & \hat{X}(z_{1,N_1-1}, z_{21}) & \cdots & \hat{X}(z_{1,N_1-1}, z_{2,N_2-1}) \end{bmatrix}, \tag{2.120}$$

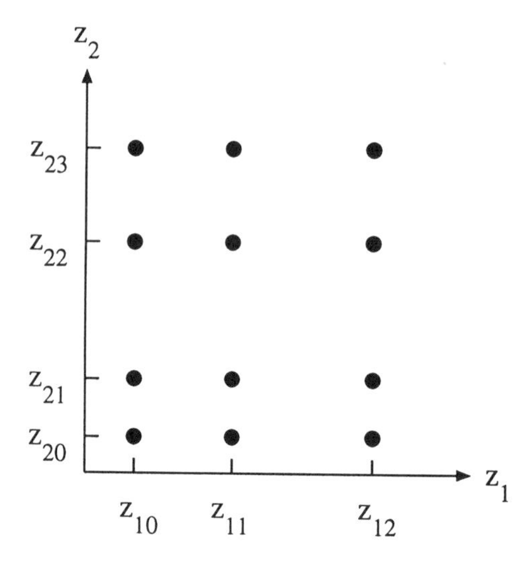

Figure 2.5. 2-D NDFT with a nonuniformly spaced rectangular grid for $N_1 = 3$, $N_2 = 4$.

$$\mathbf{X} = \begin{bmatrix} x[0,0] & x[0,1] & \cdots & x[0,N_2-1] \\ x[1,0] & x[1,1] & \cdots & x[1,N_2-1] \\ \vdots & \vdots & \ddots & \vdots \\ x[N_1-1,0] & x[N_1-1,1] & \cdots & x[N_1-1,N_2-1] \end{bmatrix}, \tag{2.121}$$

$$\mathbf{D_1} = \begin{bmatrix} 1 & z_{10}^{-1} & z_{10}^{-2} & \cdots & z_{10}^{-(N_1-1)} \\ 1 & z_{11}^{-1} & z_{11}^{-2} & \cdots & z_{11}^{-(N_1-1)} \\ \vdots & \vdots & \vdots & \ddots & \vdots \\ 1 & z_{1,N_1-1}^{-1} & z_{1,N_1-1}^{-2} & \cdots & z_{1,N_1-1}^{-(N_1-1)} \end{bmatrix}, \tag{2.122}$$

$$\mathbf{D_2} = \begin{bmatrix} 1 & z_{20}^{-1} & z_{20}^{-2} & \cdots & z_{20}^{-(N_2-1)} \\ 1 & z_{21}^{-1} & z_{21}^{-2} & \cdots & z_{21}^{-(N_2-1)} \\ \vdots & \vdots & \vdots & \ddots & \vdots \\ 1 & z_{2,N_1-1}^{-1} & z_{2,N_1-1}^{-2} & \cdots & z_{2,N_2-1}^{-(N_2-1)} \end{bmatrix}. \tag{2.123}$$

Here, $\mathbf{X}$ and $\hat{\mathbf{X}}$ are matrices of size $N_1 \times N_2$. $\mathbf{D_1}$ and $\mathbf{D_2}$ are Vandermonde matrices of sizes $N_1 \times N_1$ and $N_2 \times N_2$, respectively. The equivalent 2-D NDFT matrix $\mathbf{D}$ can be expressed in the form

$$\mathbf{D} = \mathbf{D_1} \otimes \mathbf{D_2}, \tag{2.124}$$

where $\otimes$ denotes the Kronecker product [Regalia and Mitra, 1989]. Applying a property of the Kronecker product, the determinant of $\mathbf{D}$ can be written as

$$\begin{aligned} \det(\mathbf{D}) &= \{\det(\mathbf{D_1})\}^{N_2} \otimes \{\det(\mathbf{D_2})\}^{N_1} \\ &= \prod_{i \neq j,\ i>j} ({z_{1i}}^{-1} - {z_{1j}}^{-1})^{N_2} \prod_{p \neq q,\ p>q} ({z_{2p}}^{-1} - {z_{2q}}^{-1})^{N_1}. \end{aligned} \quad (2.125)$$

Therefore, $\mathbf{D}$ is nonsingular provided $\mathbf{D_1}$ and $\mathbf{D_2}$ are nonsingular, i.e., if the points $z_{10}, z_{11}, \ldots, z_{1,N_1-1}$ are distinct, and $z_{20}, z_{21}, \ldots, z_{2,N_2-1}$ are distinct.

For this choice of sampling points, only $N_1 + N_2$ degrees of freedom are used among the $N_1 N_2$ degrees available in the 2-D NDFT. Consequently, the inverse 2-D NDFT $\mathbf{X}$ in Eq. (2.119) can be computed by solving two separate linear systems of sizes N_1 and N_2, respectively. This involves $O(N_1^3 + N_2^3)$ operations, instead of $O(N_1^3 N_2^3)$ operations in the general case.

Angelides [Angelides, 1994] has used a specific case of this sampling structure, in which the samples are placed on a nonuniform rectangular grid in the 2-D (ω_1, ω_2) plane, where $z_1 = e^{j\omega_1}$, $z_2 = e^{j\omega_2}$.

The 2-D DFT is a *special case* under this category, obtained when the points are chosen on a uniform grid in the (ω_1, ω_2) plane:

$$z_{1k_1} = e^{j\frac{2\pi}{N_1}k_1}, \qquad k_1 = 0, 1, \ldots, N_1 - 1,$$

$$z_{2k_2} = e^{j\frac{2\pi}{N_2}k_2}, \qquad k_2 = 0, 1, \ldots, N_2 - 1.$$

(2) Nonuniform sampling on parallel lines

This is a generalization of the sampling structure used in Case 1 above. For an $N_1 \times N_2$ sequence, the samples are placed on N_1 lines parallel to the z_2 axis, with N_2 points on each line. The z_1-coordinates corresponding to the N_1 lines can be chosen arbitrarily, but distinct from each other. Let these coordinates be denoted by $z_{10}, z_{11}, \ldots, z_{1,N_1-1}$. Similarly, the z_2-coordinates of the N_2 points on each line can be chosen arbitrarily, as long as they are distinct. Let the z_2-coordinates of the points on the ith line be $z_{20i}, z_{21i}, \ldots, z_{2,N_2-1,i}$. Fig. 2.6 shows an example where $N_1 = 3$, $N_2 = 4$.

In this case, the 2-D NDFT matrix can be expressed as a generalized Kronecker product [Regalia and Mitra, 1989]

$$\mathbf{D} = \{\mathbf{D_2}\} \otimes \mathbf{D_1}. \quad (2.126)$$

Here, $\mathbf{D_1}$ is an $N_1 \times N_1$ Vandermonde matrix as shown in Eq. (2.122). $\{\mathbf{D_2}\}$ denotes a set of N_1 $N_2 \times N_2$ Vandermonde matrices $\mathbf{D_{2i}}$, $i = 0, 1, \ldots, N_1 - 1$. This is represented as

$$\{\mathbf{D_2}\} = \left\{ \begin{array}{c} D_{20} \\ D_{21} \\ \vdots \\ D_{2,N_1-1} \end{array} \right\}, \quad (2.127)$$

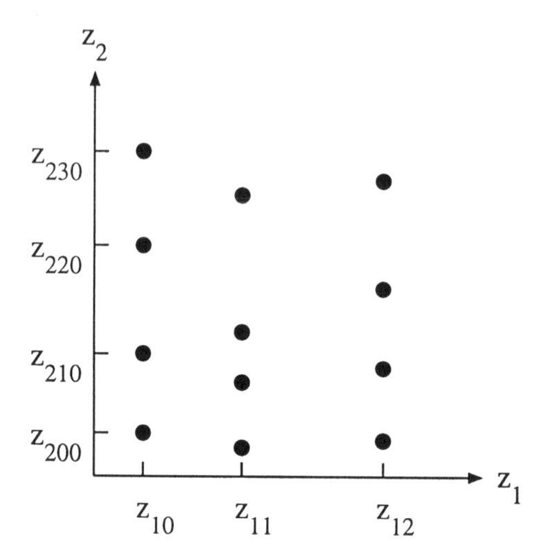

Figure 2.6. 2-D NDFT with nonuniform sampling on parallel lines (parallel to the z_2 axis) for $N_1 = 3$, $N_2 = 4$.

where

$$\mathbf{D_{2i}} = \begin{bmatrix} 1 & z_{20i}^{-1} & z_{20i}^{-2} & \cdots & z_{20i}^{-(N_2-1)} \\ 1 & z_{21i}^{-1} & z_{21i}^{-2} & \cdots & z_{21i}^{-(N_2-1)} \\ \vdots & \vdots & \vdots & \ddots & \vdots \\ 1 & z_{2,N_1-1,i}^{-1} & z_{2,N_1-1,i}^{-2} & \cdots & z_{2,N_2-1,i}^{-(N_2-1)} \end{bmatrix}, \quad i = 0, 1, \ldots, N_1 - 1. \tag{2.128}$$

Eq. (2.126) means that

$$\mathbf{D} = \begin{bmatrix} \mathbf{D_{20}} \otimes \mathbf{d_0} \\ \mathbf{D_{21}} \otimes \mathbf{d_1} \\ \vdots \\ \mathbf{D_{2,N_1-1}} \otimes \mathbf{d_{N_1-1}} \end{bmatrix}, \tag{2.129}$$

where $\mathbf{d_i}$ denotes the ith row vector of matrix $\mathbf{D_1}$.

The determinant of $\mathbf{D}$ can then be written as

$$\det(\mathbf{D}) = \{\det(\mathbf{D_1})\}^{N_2} \prod_{i=0}^{N_1-1} \det(\mathbf{D_{2i}}). \tag{2.130}$$

Therefore, $\mathbf{D}$ is nonsingular if the matrices $\mathbf{D_1}$ and $\mathbf{D_{2i}}$ are nonsingular. Alternatively, we can also place the samples on N_2 lines parallel to the z_1 axis with N_1 points on each line.

To compute the inverse 2-D NDFT, we have to solve for $x[n_1, n_2]$ from

$$\begin{aligned}\hat{X}(z_{1i}, z_{2ji}) &= \sum_{n_2=0}^{N_2-1} \sum_{n_1=0}^{N_1-1} x[n_1, n_2] z_{1i}^{-n_1} z_{2ji}^{-n_2} \\ &= \sum_{n_2=0}^{N_2-1} y[z_{1i}, n_2] z_{2ji}^{-n_2}, \qquad (2.131) \\ & \qquad i = 0, 1, \ldots, N_1 - 1, \qquad j = 0, 1, \ldots, N_2 - 1,\end{aligned}$$

where

$$y[z_{1i}, n_2] = \sum_{n_1=0}^{N_1-1} x[n_1, n_2] z_{1i}^{-n_1}. \qquad (2.132)$$

For each z_{1i}, $i = 0, 1, \ldots, N_1 - 1$, Eq. (2.131) represents a Vandermonde system of N_2 equations which can be solved to find $y[z_{1i}, n_2]$, $n_2 = 0, 1, \ldots, N_2 - 1$. For each $n_2 = 0, 1, \ldots, N_2 - 1$, Eq. (2.132) represents a Vandermonde system of N_1 equations which can be solved to obtain $x[n_1, n_2]$, $n_1 = 0, 1, \ldots, N_1 - 1$. Therefore, the inverse 2-D NDFT can be computed by solving N_1 linear systems of size N_2 each, and N_2 linear systems of size N_1 each. This requires $O(N_1 N_2^3 + N_2 N_1^3)$ operations.

Rozwood et al. [Rozwood et al., 1991] have used a specific case of this sampling structure, in which the samples were placed on parallel lines in the 2-D (ω_1, ω_2) plane.

2.6 SUMMARY

The concept of the nonuniform discrete Fourier transform, which corresponds to a nonuniform sampling of the z-transform, was introduced in this chapter. The main advantage of the NDFT is the flexibility which it provides for locating samples in the z-domain. An example was presented to demonstrate how this freedom can be utilized in spectral analysis. We showed that the problem of computing the inverse NDFT is analogous to the problem of polynomial interpolation and, therefore, can be solved by polynomial interpolation techniques. Several methods of computing the NDFT were also discussed, including the Goertzel algorithm. The subband NDFT algorithm was proposed for fast, approximate computation of the NDFT, when the signal has its energy concentrated in only a few bands of the spectrum. The concept of the NDFT was extended to higher dimensions. Having covered the preliminary basics of the NDFT, we will focus on specific applications of the NDFT in the next few chapters. The principal challenge is to find a good method of choosing the sample locations. As we shall see later, this usually depends upon the application at hand.

3 1-D FIR FILTER DESIGN USING THE NDFT

3.1 INTRODUCTION

Filters are used to select or suppress components of signals at certain frequencies. Finite-impulse-response (FIR) digital filters are of importance in many applications because they offer a number of advantages over infinite-impulse-response (IIR) filters. These advantages include the ability to attain exactly linear-phase characteristics, efficient realization through high-speed convolution using the FFT [Gold and Rader, 1969], guaranteed stability, and negligible quantization problems in implementation. In this chapter, we develop an application of the NDFT for designing 1-D FIR filters.

Commonly used techniques for FIR filter design are the windowed Fourier series approach, frequency sampling approach, and optimal minimax designs. The filters designed by the windowing and frequency sampling methods are not optimal. However, these methods are often used in practice because of their conceptual and computational simplicity. We propose a new filter design method based on nonuniform frequency sampling. The resulting filters are *nearly equal to optimal filters* designed with the same design specifications. Additionally, the *filter design times are much lower* than those required by the Parks-McClellan algorithm for optimal filter design [McClellan et al., 1973].

The outline of this chapter is as follows. In Section 3.2, we discuss the existing frequency sampling techniques for FIR filter design. The proposed NDFT-based design method is presented in Section 3.3, and is developed for designing various types of filters. In Section 3.4, we demonstrate the results obtained by this method. Design examples are shown to illustrate the effectiveness of the new method, as compared with uniform frequency sampling techniques and the Parks-McClellan algorithm.

3.2 EXISTING METHODS FOR FREQUENCY SAMPLING DESIGN

In the traditional frequency sampling approach to FIR filter design, the desired frequency response is sampled in the frequency range $0 \leq \omega \leq 2\pi$, at N *equally spaced* frequencies, where N is the filter length. Since the frequency samples are simply the elements of the DFT of the filter impulse response, an N-point inverse DFT is used to compute the impulse response, i.e., the filter coefficients. The frequency response $H(e^{j\omega})$ of the resulting filter is, therefore, an interpolation between the frequency samples. Although exact frequency response values are obtained at the specified frequencies, the approximation error between the interpolated response and the desired response can be unacceptably large at intermediate frequencies. The interpolated response has *very large ripples* in regions where there are sudden transitions in the desired response, e.g., near the band edges of band-selective filters. This occurs because the desired frequency response is usually piecewise constant, with step-like transitions between successive bands. Such a frequency response corresponds to an impulse response of infinite length. Thus, frequency sampling produces an FIR filter impulse response which is a time-aliased version of this infinite sequence. To improve the filter characteristics, a transition band has been introduced in the desired frequency response.

The frequency sampling approach to FIR filter design was introduced originally by Gold and Jordan [Gold and Jordan, Jr., 1969], and developed further by Rabiner et al. [Rabiner et al., 1970]. Here, a number of samples in the transition band are *varied in amplitude* so as to minimize the maximum deviation from the desired response over some frequency range of interest. Linear programming is used to optimize the values of these variable samples. However, since this procedure is computationally intensive, the number of frequency samples which can be varied simultaneously must be small. This limits the length of filters that can be designed in practice. Besides, this method does not provide good control over typical filter specifications such as the band edges and peak ripple. An improved uniform frequency sampling technique, introduced by Jarske et al. [Jarske et al., 1988; Lightstone et al., 1994], uses a desired response that is a simple analytic function with a few degrees of freedom. The parameters controlling the desired response are chosen in such a way that the response can be realized almost exactly by an FIR filter. This *analytic function* is sampled uniformly to obtain frequency samples, whose inverse DFT then gives the filter coefficients.

A brief discussion of the technique of *nonuniform frequency sampling* was included in the paper by Rabiner et al. [Rabiner et al., 1970]. However, they considered lowpass filter design examples using only two sets of nonuniform data. The first set consisted of uniformly spaced samples with an extra sample placed in the transition band. The second set had two extra samples in the transition band. The uniform samples had values of unity in the passband and zeros in the stopband. In both these cases, the locations of the transition samples were kept fixed. The design criterion adopted was to choose optimum values for the nonuniform samples so as to minimize the peak ripple in the stopband. They compared the peak ripples of the resulting filters with those obtained by uniform frequency sampling. Although lower peak ripples were obtained in the first case, higher values were obtained in the second case. Thus, the results obtained were not entirely encouraging and led the authors to stress that more work would have to be done to fully utilize the advantages of nonuniform frequency sampling. Note that we can apply the Lagrange interpolation formula in Eq. (2.11) to express the z-transform of the filter $H(z)$ in the form

$$H(z) = \sum_{k=0}^{N-1} \frac{H_k \sum_{i=0}^{N-1} b_i z^{-i}}{\alpha_k (1 - z_k z^{-1})}, \tag{3.1}$$

where

$$H_k = H(z_k), \qquad k = 0, 1, \ldots, N-1,$$

$$\alpha_k = \prod_{i \neq k} (1 - z_i z_k^{-1})$$

and the coefficients b_i satisfy

$$\sum_{i=0}^{N-1} b_i z^{-i} = \prod_{i=0}^{N-1} (1 - z_i z^{-1}).$$

Although $H(z)$ is a linear function of the sample values H_k, it is a nonlinear function of the sample locations z_k. Thus, we have to resort to complex nonlinear optimization techniques if we want to choose the values as well as the locations of the frequency samples, so as to minimize the peak ripple in some band. In a later effort on nonuniform frequency sampling, Angelides et al. [Angelides and Diamessis, 1994] used a Newton-type polynomial with complex conjugate coefficients for interpolating the desired complex frequency response of a real-coefficient FIR filter.

3.3 PROPOSED NONUNIFORM FREQUENCY SAMPLING DESIGN

In the proposed FIR filter design method, the desired frequency response is sampled at N nonuniformly spaced points on the unit circle in the z-plane [Bagchi and Mitra, 1996a]. An N-point inverse NDFT of these frequency samples gives the filter coefficients. As discussed earlier in Section 2.1, this procedure is mathematically equivalent to computing the coefficients of an interpolating polynomial of order $N-1$, that has the desired values at the N specified frequency locations [Rabiner et al., 1970]. Besides, the representation of the filter polynomial $H(z)$ by various interpolation methods, such as the Lagrange and Newton forms, leads to various possible filter structures with different merits [Rabiner et al., 1970; Schüssler, 1972].

We address the problem of nonuniform frequency sampling design by considering the two major issues involved: (a) the generation of the desired frequency response for the given filter specifications, and (b) the choice of the frequency-sample locations. The aim is to obtain a filter whose interpolated frequency response is nearly equiripple in each band. The approaches used for these two issues are outlined here:

(a) **Generation of the desired frequency response**: Given the filter specifications, we construct the desired response by using a separate *analytic function* for each frequency band. The equiripple nature of the response is obtained by using Chebyshev polynomials and a suitable transformation to map the domain of the polynomial to the frequency axis. Each function involves several parameters which are found by imposing appropriate constraints so as to obtain the desired mapping.

(b) **Choice of the frequency-sample locations**: In addressing this issue, we must note that there is no good analytic approximation for the frequency response in the transition band. We have found that the best locations for the frequency samples are the *extrema* of the desired equiripple response, as shown later in this chapter. Since the desired response is approximated by analytic functions, the extremal frequencies can be expressed in closed form. In addition, since the total number of extrema is related to the filter length, we do not need to place any extra samples in the transition band. However, in special cases (e.g., for designing filters with very wide transition bands), we can place a few transition samples to constrain any possible overshoot within this band.

Once we have addressed these issues, we can refer to the NDFT formulation in Eq. (2.1). Samples of the analytic functions generated in Step (a) are used to construct the NDFT vector $\mathbf{X}$. The NDFT matrix $\mathbf{D}$ is constructed from the frequency-sample locations found in Step (b). The filter impulse response $\mathbf{x}$ is then obtained by solving this linear system of equations. Note that all known symmetries in the filter impulse response can be utilized to reduce the number of independent filter coefficients. This leads to a smaller linear system, and, therefore, to reduced design time.

We can use the proposed method to design various filters (of length N), such as lowpass filters (Type I if N is odd, Type II if N is even), highpass filters (Type I if N is odd, Type IV if N is even), bandpass filters (Types I or III if N is odd, Types II or IV if N is even), and Mth-band filters such as third-band filters (Type I). We now describe the design method in detail by considering a few of these filters.

3.3.1 *Lowpass Filter Design (Type I)*

Let us consider the design of a linear-phase lowpass filter of Type I. The filter has a real and symmetric impulse response

$$h[n] = h[N-1-n], \qquad n = 0, 1, \ldots, N-1, \tag{3.2}$$

where the filter length N is an odd integer. The frequency response of the filter is

$$H(e^{j\omega}) = \sum_{n=0}^{N-1} h[n]e^{-j\omega n}. \tag{3.3}$$

By applying the symmetry condition in Eq. (3.2), this can be expressed in the form [Mitra, 1998]

$$H(e^{j\omega}) = A(\omega)e^{-j\omega(N-1)/2}, \tag{3.4}$$

where the amplitude function $A(\omega)$ is a real, even, periodic function of ω, given by

$$A(\omega) = \sum_{k=0}^{(N-1)/2} a[k]\cos\omega k, \tag{3.5}$$

and

$$\begin{aligned} a[0] &= h\left[(N-1)/2\right], \\ a[k] &= 2h[(N-1)/2-k], \qquad k = 1, 2, \ldots, (N-1)/2. \end{aligned} \tag{3.6}$$

Let the filter have its passband edge at ω_p, stopband edge at ω_s, and peak ripples δ_p and δ_s in the passband and stopband, respectively. We proceed with the design by considering the two issues involved:

(a) Generation of the desired frequency response

The real-valued amplitude response $A(\omega)$ is represented by analytic functions as follows [Bagchi and Mitra, 1996a]:

$$A(\omega) = \begin{cases} H_p(\omega), & 0 \le \omega \le \omega_p, \\ H_s(\omega), & \omega_s \le \omega \le \pi, \end{cases} \tag{3.7}$$

where

$$H_p(\omega) = 1 - \delta_p T_P(X_p(\omega)) \tag{3.8}$$

and

$$H_s(\omega) = -\delta_s T_S(X_s(\omega)), \tag{3.9}$$

with $T_M(\cdot)$ denoting a Chebyshev polynomial of order M, defined as

$$T_M(x) = \begin{cases} \cos(M\cos^{-1}(x)), & -1 \le x \le 1, \\ \cosh(M\cosh^{-1}(x)), & \text{otherwise.} \end{cases} \tag{3.10}$$

Note that $T_M(x)$ is equiripple in the range $-1 \leq x \leq 1$ and monotone outside this range. Depending on whether n is even or odd, $T_M(x)$ is an even or odd function of x, respectively. The integers P and S are given by

$$P = N_p \tag{3.11}$$

and

$$S = N_s, \tag{3.12}$$

where N_p equals the number of extrema in the passband $0 \leq \omega \leq \omega_p$, and N_s equals the number of extrema in the stopband $\omega_s \leq \omega \leq \pi$. The functions, $X_p(\omega)$ and $X_s(\omega)$, are needed to map the equiripple interval $-1 \leq x \leq 1$ of $T_P(x)$ and $T_S(x)$ to the passband and stopband, respectively. This mapping is obtained by using the transformations,

$$X_p(\omega) = A\cos(a\omega + b) + B, \tag{3.13}$$

$$X_s(\omega) = C\cos(c\omega + d) + D. \tag{3.14}$$

The values for the eight parameters, A, B, C, D, a, b, c, d, are obtained by imposing appropriate constraints on the functions, $H_p(\omega)$ and $H_s(\omega)$ [Bagchi and Mitra, 1996a]. The desired response is generated by the following eight steps.

Step 1

From Eq. (3.5), since $A(\omega)$ is an even function of ω, we have

$$A(\omega) = A(-\omega). \tag{3.15}$$

Therefore, from Eq. (3.7), $H_p(\omega)$ should be symmetric with respect to $\omega = 0$; that is,

$$H_p(\omega) = H_p(-\omega).$$

If we substitute for $H_p(\omega)$ from Eq. (3.8), we have

$$X_p(\omega) = X_p(-\omega).$$

Substituting for $X_p(\omega)$ from Eq. (3.13), we obtain

$$\cos(a\omega + b) = \cos(-a\omega + b),$$

or

$$b = \pi. \tag{3.16}$$

Step 2

From Eq. (3.5), we observe that

$$A(\pi + \omega) = A(\pi - \omega). \tag{3.17}$$

Therefore, from Eq. (3.7), $H_s(\omega)$ should be symmetric with respect to $\omega = \pi$; that is,

$$H_s(\pi + \omega) = H_s(\pi - \omega).$$

If we substitute for $H_s(\omega)$ from Eq. (3.9), we have

$$X_s(\pi + \omega) = X_s(\pi - \omega).$$

Substituting for $X_s(\omega)$ from Eq. (3.14), we obtain

$$\cos(c\pi + c\omega + d) = \cos(c\pi - c\omega + d),$$

or

$$c\pi + d = \pi,$$

or

$$d = \pi(1 - c). \tag{3.18}$$

Step 3

Since ω_p is the passband edge, an optimal equiripple response must have an alternation at ω_p [Oppenheim and Schafer, 1989]. Thus, we impose the condition

$$H_p(\omega_p) = 1 - \delta_p. \tag{3.19}$$

The expression for $H_p(\omega)$ in Eq. (3.8) implies that

$$T_P(X_p(\omega_p)) = 1.$$

Using the definition of $T_P(\cdot)$ given in Eq. (3.10), we have

$$X_p(\omega_p) = 1.$$

If we substitute for $X_p(\omega)$ from Eq. (3.13), we obtain

$$A\cos(a\omega_p + b) + B = 1.$$

Substituting for b from Eq. (3.16), we have

$$a = \frac{1}{\omega_p} \cos^{-1}\left(\frac{B-1}{A}\right). \tag{3.20}$$

Step 4

Since ω_s is the stopband edge, an optimal equiripple response must have an alternation at ω_s [Oppenheim and Schafer, 1989]. Thus,

$$H_s(\omega_s) = \delta_s. \tag{3.21}$$

The expression for $H_s(\omega)$ in Eq. (3.9) implies that

$$T_S(X_s(\omega_s)) = 1.$$

Using the definition of $T_S(\cdot)$ given in Eq. (3.10), we have

$$X_s(\omega_s) = 1.$$

If we substitute for $X_s(\omega)$ from Eq. (3.14), we obtain

$$C\cos(c\omega_s + d) + D = 1.$$

Substituting for d from Eq. (3.18), we have

$$c = \frac{1}{(\omega_s - \pi)} \cos^{-1}\left(\frac{D-1}{C}\right). \tag{3.22}$$

Step 5

We constrain $H_p(\omega)$ so that an extremum occurs at $\omega = 0$. Therefore,

$$H_p(0) = \begin{cases} 1 + \delta_p, & P = \text{odd}, \\ 1 - \delta_p, & P = \text{even}. \end{cases} \tag{3.23}$$

Using the definition of $H_p(\omega)$ in Eq. (3.8), we obtain the condition,

$$T_P(X_p(0)) = \begin{cases} -1, & P = \text{odd}, \\ 1, & P = \text{even}. \end{cases}$$

From Eq. (3.10), we have

$$X_p(0) = -1.$$

Substituting for $X_p(\omega)$ from Eq. (3.13), and using the value of b found in Eq. (3.16), we obtain

$$B = A - 1. \tag{3.24}$$

Step 6

We constrain $H_s(\omega)$ so that an extremum occurs at $\omega = \pi$. Thus, we have

$$H_s(\pi) = \begin{cases} -\delta_s, & S = \text{odd}, \\ \delta_s, & S = \text{even}. \end{cases} \tag{3.25}$$

Using the definition of $H_s(\omega)$ in Eq. (3.9), we obtain the condition,

$$T_S(X_s(\pi)) = \begin{cases} -1, & S = \text{odd}, \\ 1, & S = \text{even}, \end{cases}$$

or

$$X_s(\pi) = -1.$$

Substituting for $X_s(\omega)$ from Eq. (3.14), we obtain

$$D = C - 1. \tag{3.26}$$

Note that as a result of the conditions in Steps 5 and 6, the filters designed using this method always have extrema at $\omega = 0$ and $\omega = \pi$. Some optimal equiripple filters of Type I have an extremum at only one of these frequencies. However, the filters designed by our method have nearly equiripple frequency responses, comparable to optimal filter responses.

Steps 1–6 give us six equations connecting the eight unknown parameters. So as to solve for these parameters, we obtain two more equations by requiring that $H_p(\omega)$ and $H_s(\omega)$ stay between $-\delta_s$ and $(1 + \delta_p)$ at all frequencies. This leads to the following steps.

Step 7

$$\min(H_p(\omega)) = -\delta_s \tag{3.27}$$

Using Eqs. (3.8), (3.10) and (3.13), we obtain

$$1 - \delta_p T_P(A + B) = -\delta_s.$$

We substitute for B from Eq. (3.24), and find that

$$A = \frac{1}{2}\left\{T_P{}^{-1}\left(\frac{1+\delta_s}{\delta_p}\right) + 1\right\}, \tag{3.28}$$

where $T_M{}^{-1}(\cdot)$ denotes an inverse Chebyshev function of order M.

Step 8

$$\max(H_s(\omega)) = 1 + \delta_p \tag{3.29}$$

Using Eqs. (3.9), (3.10) and (3.14), we obtain

$$\delta_s T_S(C + D) = 1 + \delta_p.$$

We substitute for D from Eq. (3.26), to obtain

$$C = \frac{1}{2}\left\{T_S{}^{-1}\left(\frac{1+\delta_p}{\delta_s}\right) + 1\right\}. \tag{3.30}$$

Thus, we have found closed-form expressions for the eight parameters, A, B, C, D, a, b, c, d, as shown in Eqs. (3.28), (3.24), (3.30), (3.26), (3.20), (3.16), (3.22) and (3.18).

Given the filter specifications, N, ω_p, ω_s and $k = \delta_p/\delta_s$, we estimate the ripple sizes δ_p and δ_s from [Kaiser, 1974]

$$N = \frac{-10\log_{10}(\delta_p\delta_s) - 13}{2.324(\omega_s - \omega_p)} + 1, \tag{3.31}$$

which gives

$$\delta_p = \sqrt{k}\ 10^{-0.1162(\omega_s-\omega_p)(N-1)-0.65} \tag{3.32}$$

and

$$\delta_s = k/\delta_p. \tag{3.33}$$

Alternatively, if we are given ω_p, ω_s, δ_p and δ_s, we can estimate the required filter length N from Eq. (3.31). The Chebyshev polynomial orders P and S are then determined. Since they correspond to the number of extrema in the passband and stopband, they are found by weighting them proportionately to the sizes of the passband and stopband such that $P + S = (N - 1)/2$. Now, $P+S$ equals the total number of alternations in the filter response over the range $0 \leq \omega \leq \pi$, excluding the ones at the band edges, ω_p and ω_s. The alternation theorem states that an optimum Type I filter must have a minimum of $(L+2)$ alternations, where $L = (N-1)/2$ [Oppenheim and Schafer, 1989]. Thus, we obtain

$$P + S = L = \frac{N-1}{2}. \tag{3.34}$$

Note that a Type I filter can have a maximum of $(L+3)$ alternations [Oppenheim and Schafer, 1989]. For this extraripple filter, we have $P+S = (N+1)/2$.

(b) Choice of the frequency-sample locations

Since the impulse response of the filter is symmetric, as shown in Eq. (3.2), the number of independent filter coefficients is given by

$$N_i = \frac{N+1}{2}. \tag{3.35}$$

Thus, we need only N_i samples located in the range $0 \leq \omega \leq \pi$. These samples are placed at the extrema of the desired response which has been approximated by analytic functions in Step (a).

The P extrema of $H_p(\omega)$ occur when

$$H_p(\omega) = 1 \pm \delta_p, \tag{3.36}$$

or

$$T_P(X_p(\omega)) = \pm 1,$$

or

$$X_p(\omega) = \cos\left(\frac{k\pi}{P}\right), \qquad k = 1, 2, \ldots, N_p. \tag{3.37}$$

Using the definition for $X_p(\omega)$ in Eq. (3.13), we obtain the following expression for the extrema in the passband:

$$\omega_k^{(p)} = \frac{1}{a}\left\{\cos^{-1}\left[\frac{\cos\left(\frac{k\pi}{P}\right) - B}{A}\right] - b\right\}, \qquad k = 1, 2, \ldots, N_p. \tag{3.38}$$

Similarly, the S extrema of $H_s(\omega)$ occur when

$$H_s(\omega) = \pm\delta_s. \tag{3.39}$$

This leads to the following expression for the extrema in the stopband:

$$\omega_k^{(s)} = \frac{1}{c}\left\{\cos^{-1}\left[\frac{\cos\left(\frac{k\pi}{S}\right) - D}{C}\right] - d\right\}, \qquad k = 1, 2, \ldots, N_s. \tag{3.40}$$

From Eqs. (3.34) and (3.35), we observe that $P + S = N_i - 1$. Therefore, we need one more sample besides those at the $P + S$ extrema. This sample is placed either at the passband edge ω_p, or at the stopband edge ω_s.

Thus, we sample the functions generated in Step (a) at the locations obtained in Step (b), and then solve for the N_i independent filter coefficients. We can also design highpass filters using the proposed method with minor modifications, by designing the function $H_s(\omega)$ for the low-frequency band and $H_p(\omega)$ for the high-frequency band.

Example 3.1

In this example, we illustrate the method of Type I lowpass filter design discussed in Section 3.3.1. The filter specifications are:

Filter length $N = 37$, Ripple ratio $\delta_p/\delta_s = 1$,
Passband edge $\omega_p = 0.3\pi$, Stopband edge $\omega_s = 0.4\pi$.

The orders of the Chebyshev polynomials for the passband and stopband are: $P = 6$, $S = 12$. Fig. 3.1(a) shows a plot of $T_P(x)$. The equiripple region $-1 \leq x \leq 1$ of $T_P(x)$ is mapped to the passband by the transformation $X_p(\omega)$, which is plotted in Fig. 3.1(b). Fig. 3.1(c) shows the functions, $H_p(\omega)$ and $H_s(\omega)$, which approximate the passband and stopband of the desired response; these functions are sampled at the locations marked on this figure. The resulting filter has a frequency response that is nearly equiripple, as shown in Fig. 3.1(d).

3.3.2 Half-band Filter Design

Half-band FIR filters belong to a special class of Type I FIR filters. The design procedure discussed for Type I lowpass filters in Section 3.3.1 is further simplified because every other sample in the half-band impulse response is zero:

$$h[n] = \begin{cases} 0, & n - \frac{N-1}{2} = \text{even and nonzero}, \\ 0.5, & n = \frac{N-1}{2}. \end{cases} \tag{3.41}$$

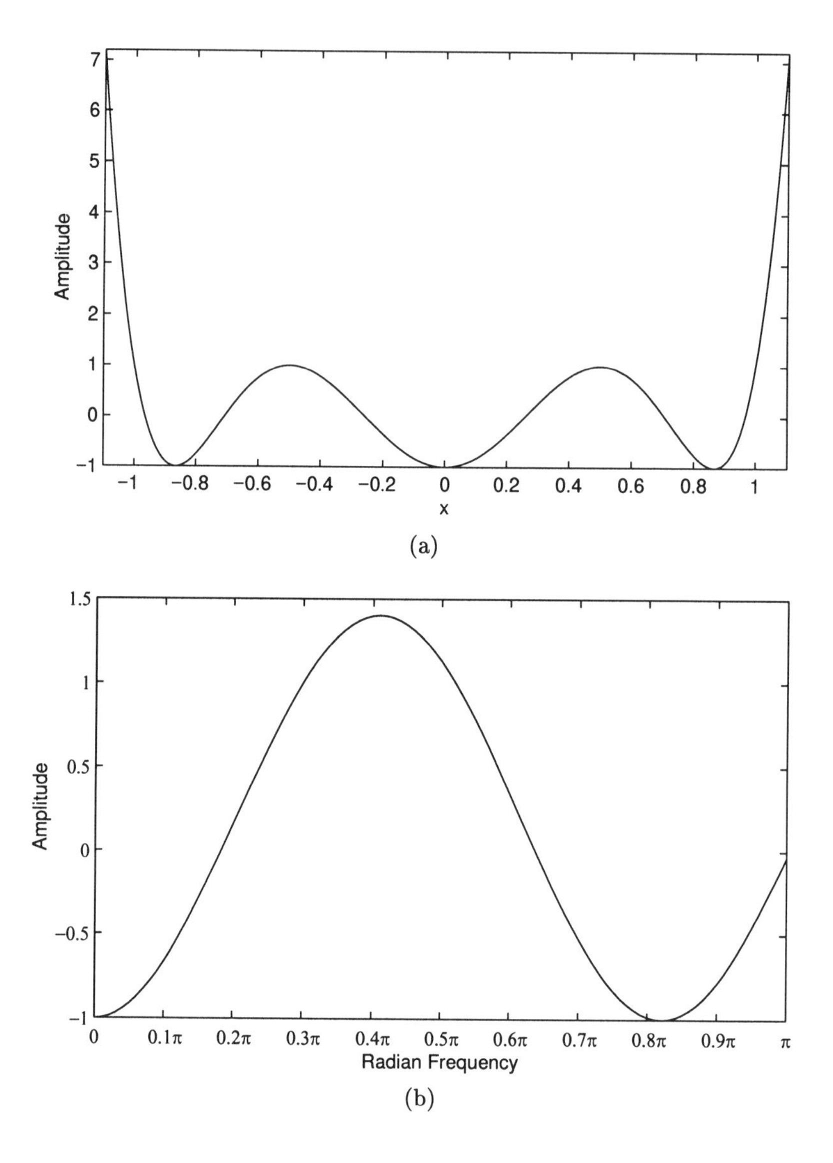

Figure 3.1. Type I lowpass filter design example. (a) Plot of Chebyshev polynomial $T_P(x)$ as a function of x. (b) Plot of $X_p(\omega)$ with normalized frequency $(\omega/2\pi)$.

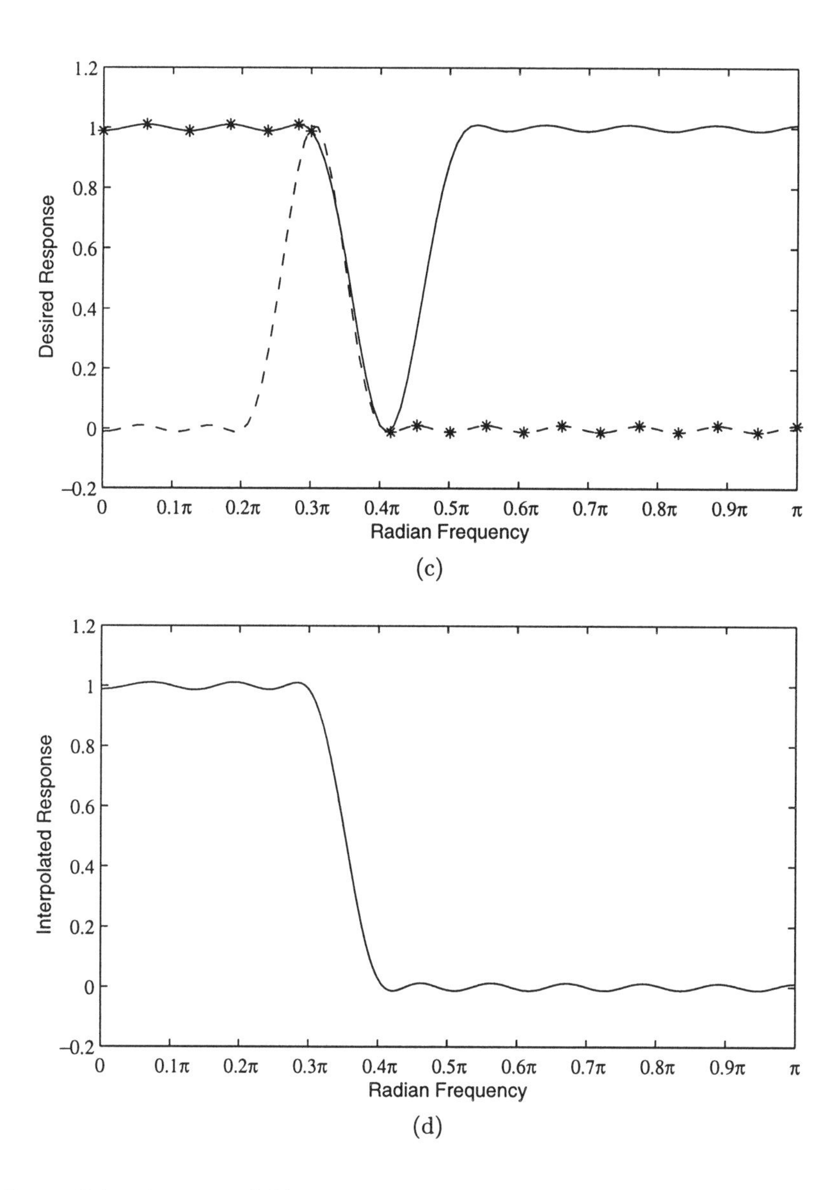

Figure 3.1. (continued) (c) Desired response approximated by $H_p(\omega)$ (solid line) in the passband, and $H_s(\omega)$ (dashed line) in the stopband. These functions are sampled at the locations denoted by "$*$". (d) Interpolated response of resulting lowpass filter.

This leads to the following symmetry in the frequency response of the filter:

$$H(e^{j\omega}) + H(e^{j(\pi-\omega)}) = 1. \tag{3.42}$$

In addition, the band edges and peak ripples in the passband and stopband are related as follows:

$$\omega_p + \omega_s = \pi, \tag{3.43}$$

$$\delta_p = \delta_s. \tag{3.44}$$

From the half-band condition in Eq. (3.41), we find that the number of independent filter coefficients is given by

$$N_i = \left\lceil \frac{N-1}{4} \right\rceil, \tag{3.45}$$

where $\lceil x \rceil$ denotes the integer closest to and larger than x. The desired response is generated as discussed in Section 3.3.1. From the symmetry condition in Eq. (3.42), it follows that the orders of the Chebyshev polynomials approximating the passband and stopband are equal. Thus,

$$P = S = N_i. \tag{3.46}$$

The symmetry between the passband and stopband also results in fewer parameters controlling the functions $H_p(\omega)$ and $H_s(\omega)$, since

$$A = C, \tag{3.47}$$

$$B = D, \tag{3.48}$$

$$a = c. \tag{3.49}$$

We need to generate only the four parameters, A, B, a, and b, given by Eqs. (3.28), (3.24), (3.20) and (3.16) from the given filter specifications. Then, we obtain N_i samples at the P extrema of the passband function $H_p(\omega)$, as given in Eq. (3.38). This illustrates the flexibility of our design method. Besides, the design time is reduced nearly by *half*, as compared with that required by a general symmetric filter of the same length.

Example 3.2

Consider the design a half-band lowpass filter with the following specifications:

Filter length $N = 33$,
Passband edge $\omega_p = 0.46\pi$, Stopband edge $\omega_s = 0.54\pi$.

Since the symmetry in the half-band frequency response allows us to place all the samples in the passband, we need to generate the desired response for the passband only. There are only eight independent filter coefficients. Note that this is nearly half of 17, which is the number of independent coefficients in a general symmetric filter of length 33. Thus, the order of the Chebyshev polynomial for the passband is $P = 8$. Fig. 3.2(a) shows the function $H_p(\omega)$, which approximates the passband of the desired response, and the sample locations. The frequency response of the filter obtained is shown in Fig. 3.2(b).

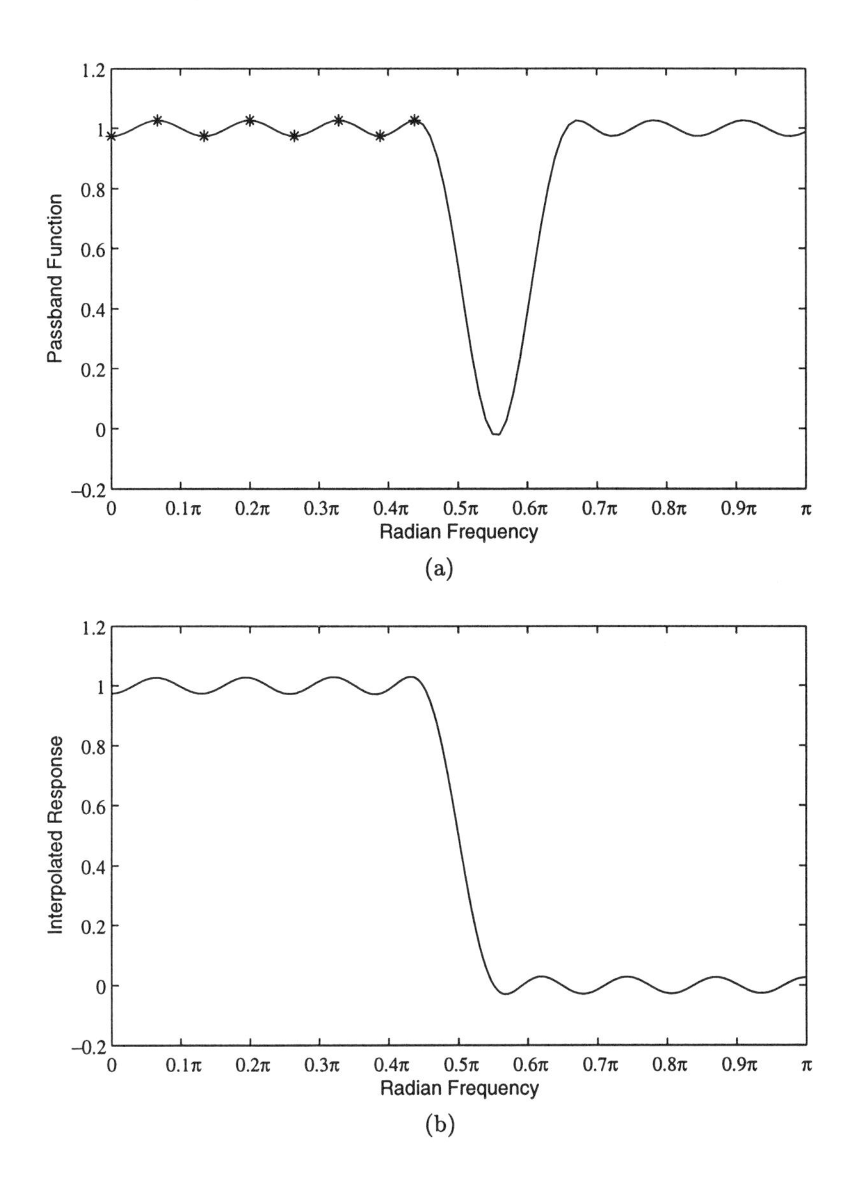

Figure 3.2. Half-band lowpass filter design example. (a) Plot of $H_p(\omega)$, which approximates the desired response in the passband. The sample locations are denoted by "$*$". (b) Interpolated response of resulting half-band filter.

3.3.3 Lowpass Filter Design (Type II)

We shall now modify the method discussed for Type I lowpass filters in Section 3.3.1 so as to design a linear-phase lowpass filter of Type II. Such a filter has a symmetric impulse response as shown in Eq. (3.2), where the filter length N is an even integer. The frequency response of the filter can be expressed as [Mitra, 1998]

$$H(e^{j\omega}) = B(\omega)e^{-j\omega(N-1)/2}, \tag{3.50}$$

where $B(\omega)$ is a real, even, periodic function of ω, given by

$$B(\omega) = \sum_{k=1}^{N/2} b[k]\cos[\omega(k - \frac{1}{2})], \tag{3.51}$$

and

$$b[k] = 2h[N/2 - k], \qquad k = 1, 2, \ldots, N/2. \tag{3.52}$$

The real-valued amplitude response $B(\omega)$ is represented by

$$B(\omega) = \begin{cases} H_p(\omega) = 1 - \delta_p T_P(X_p(\omega)), & 0 \leq \omega \leq \omega_p, \\ H_s(\omega) = -\delta_s T_S(X_s(\omega)), & \omega_s \leq \omega \leq \pi. \end{cases} \tag{3.53}$$

In this case, the orders of the Chebyshev polynomials in the passband and stopband are given by

$$P = N_p \tag{3.54}$$

and

$$S = 2N_s + 1, \tag{3.55}$$

where N_p and N_s equal the number of extrema in the passband and stopband regions, respectively. Eq. (3.55) arises because the Type II response $B(\omega)$ has a zero at $\omega = \pi$, and is antisymmetric about this point in frequency. Therefore, the Chebyshev polynomial $T_S(x)$ must also be antisymmetric, i.e., its order S must be odd, as shown in Eq. (3.55). The equiripple interval $0 \leq x \leq 1$ of $T_S(x)$ is mapped by the function $X_s(\omega)$ to the stopband $\omega_s \leq \omega \leq \pi$. As discussed in Section 3.3.1, the equiripple interval $-1 \leq x \leq 1$ of $T_P(x)$ is mapped to the passband $0 \leq \omega \leq \omega_p$ by the function $X_p(\omega)$. This mapping is obtained by choosing

$$X_p(\omega) = A\cos(a\omega + b) + B \tag{3.56}$$

and

$$X_s(\omega) = C\cos(c\omega + d) + D. \tag{3.57}$$

We now obtain closed-form expressions for the eight parameters, A, B, C, D, a, b, c, d, by imposing appropriate constraints on the functions, $H_p(\omega)$ and $H_s(\omega)$. This is shown in the following eight steps.

Step 1

Since $B(\omega)$ is an even function of ω, $H_p(\omega)$ should be symmetric with respect to $\omega = 0$; that is,

$$H_p(\omega) = H_p(-\omega). \tag{3.58}$$

Using Eqs. (3.53) and (3.56), we obtain

$$b = \pi. \tag{3.59}$$

Step 2

From Eq. (3.51), we note that

$$B(\pi + \omega) = -B(\pi - \omega). \tag{3.60}$$

Therefore, from Eq. (3.53), $H_s(\omega)$ should be anti-symmetric with respect to $\omega = \pi$,

$$H_s(\pi + \omega) = -H_s(\pi - \omega).$$

Substituting for $H_s(\omega)$ from Eq. (3.53), we have

$$T_S(X_s(\pi + \omega)) = -T_S(X_s(\pi - \omega)).$$

Since S is odd, this implies that

$$X_s(\pi + \omega) = -X_s(\pi - \omega).$$

Substituting for $X_s(\omega)$ from Eq. (3.57), we obtain

$$C\cos(c\pi + c\omega + d) + D = -C\cos(c\pi - c\omega + d) - D.$$

Let us assume that

$$D = 0. \tag{3.61}$$

In step 6, we shall see that this assumption is justified. Thus, we have

$$c\pi + d = \pi/2,$$

or

$$d = \frac{\pi}{2}(1 - 2c). \tag{3.62}$$

Step 3

Since an optimal equiripple response must have an alternation at the passband edge ω_p, we impose the condition

$$H_p(\omega_p) = 1 - \delta_p. \tag{3.63}$$

Using Eqs. (3.53), (3.56) and (3.59), we obtain

$$a = \frac{1}{\omega_p} \cos^{-1}\left(\frac{B-1}{A}\right). \tag{3.64}$$

Step 4

Similarly, since an optimal equiripple response must have an alternation at the stopband edge ω_s,

$$H_s(\omega_s) = \delta_s. \tag{3.65}$$

The definition of $H_s(\omega)$ in Eq. (3.53) implies that

$$T_S(X_s(\omega_s)) = 1,$$

or

$$X_s(\omega_s) = 1.$$

Substituting for $X_s(\omega)$ from Eq. (3.57), we have

$$C\cos(c\omega_s + d) + D = 0.$$

By using the expressions for D and d found in Eqs. (3.61) and (3.62), we can simplify this to obtain

$$c = \frac{1}{(\pi - \omega_s)} \sin^{-1}\left(\frac{1}{C}\right). \tag{3.66}$$

Step 5

We constrain $H_p(\omega)$ so that an extremum occurs at $\omega = 0$. Thus, we have

$$H_p(0) = \begin{cases} 1+\delta_p, & P = \text{odd}, \\ 1-\delta_p, & P = \text{even}. \end{cases} \tag{3.67}$$

Using the definitions of $H_p(\omega)$ and $X_p(\omega)$ found in Eqs. (3.53) and (3.56), we can simplify this to obtain

$$B = A - 1. \tag{3.68}$$

Step 6

Since the Type II response $B(\omega)$ has a zero at $\omega = \pi$, we must have

$$H_s(\pi) = 0. \tag{3.69}$$

Using the definition of $H_s(\omega)$ in Eq. (3.53), we obtain

$$T_S(X_s(\pi)) = 0.$$

Since S is an odd integer, this implies that

$$X_s(\pi) = 0.$$

Substituting for $X_s(\omega)$ from Eq. (3.57), we obtain

$$C\cos(c\pi + d) + D = 0. \tag{3.70}$$

From Eqs. (3.61) and (3.62), we infer that this condition is indeed satisfied by the assumption $D = 0$ in Step 2. Thus, the assumption is justified.

Step 7

$$\min(H_p(\omega)) = -\delta_s \tag{3.71}$$

Using Eqs. (3.53), (3.56) and (3.68), we obtain

$$A = \frac{1}{2}\left\{ T_P{}^{-1}\left(\frac{1+\delta_s}{\delta_p}\right) + 1 \right\}, \tag{3.72}$$

Step 8

$$\max(H_s(\omega)) = 1 + \delta_p \tag{3.73}$$

Using Eqs. (3.53), (3.57) and (3.61), we obtain

$$C = T_S{}^{-1}\left(\frac{1+\delta_p}{\delta_s}\right). \tag{3.74}$$

Thus, we have found closed-form expressions for the eight parameters, A, B, C, D, a, b, c, d, as shown in Eqs. (3.72), (3.68), (3.74), (3.61), (3.64), (3.59), (3.66) and (3.62).

In the Type II case, the Chebyshev polynomial orders P and S are determined from Eqs. (3.54) and (3.55). N_p and N_s are first found by weighting them proportionally to the sizes of the passband and stopband, such that $N_p + N_s = (N/2) - 1$. Note that $N_p + N_s$ equals the total number of alternations in the frequency response over the range $0 \leq \omega \leq \pi$, excluding the ones at the band edges, ω_p and ω_s. The alternation theorem states that an optimum Type II filter must have a minimum of $(L + 2)$ alternations, where $L = (N/2) - 1$ [Oppenheim and Schafer, 1989]. Thus, we have

$$N_p + N_s = L = \frac{N}{2} - 1. \tag{3.75}$$

Due to the symmetry in the impulse response, the number of independent filter coefficients in a Type II filter is given by

$$N_i = \frac{N}{2}. \tag{3.76}$$

Thus, we place N_i samples at the extrema of the desired response in the frequency range $0 \leq \omega \leq \pi$, as shown in Eqs. (3.38) and (3.40). Since $P + S = N_i - 1$, we need one more sample which is placed at either ω_p or ω_s.

Example 3.3

Let us design a Type II lowpass filter using the method described in Section 3.3.3. The filter specifications are:

Filter length $N = 28$, Ripple ratio $\delta_p/\delta_s = 1$,
Passband edge $\omega_p = 0.3\pi$, Stopband edge $\omega_s = 0.4\pi$.

The number of extrema in the passband and stopband are: $N_p = 5$, $N_s = 8$. Thus, the Chebyshev polynomial orders are: $P = 5$, $S = 17$. As shown in Fig. 3.3(a), the Chebyshev polynomial $T_S(x)$ is antisymmetric with respect to the point $x = 0$. Since the filter must have a zero at $\omega = \pi$, the equiripple region $0 \leq x \leq 1$ of $T_S(x)$ is mapped to the stopband by the function $X_s(\omega)$, which is shown in Fig. 3.3(b). The functions $H_p(\omega)$ and $H_s(\omega)$ are plotted in Fig. 3.3(c), along with the sample locations. The frequency response of the resulting filter is shown in Fig. 3.3(d).

3.3.4 Bandpass Filter Design

Without loss of generality, we assume that the bandpass filter is of Type I. Then its frequency response can be expressed as shown in Eq. (3.4). Let the filter have a passband $\omega_{p_1} \leq \omega \leq \omega_{p_2}$, and two stopbands $0 \leq \omega \leq \omega_{s_1}$ and $\omega_{s_2} \leq \omega \leq \pi$, with peak ripples δ_p, δ_{s_1} and δ_{s_2}, respectively.

As in lowpass filter design, the desired frequency response is represented by analytic functions, one for each band. The real-valued amplitude response $A(\omega)$ is represented by

$$A(\omega) = \begin{cases} H_{s_1}(\omega) = -\delta_{s_1} T_{S_1}(X_{s_1}(\omega)), & 0 \leq \omega \leq \omega_{s_1}, \\ H_p(\omega) = 1 - \delta_p T_P(X_p(\omega)), & \omega_{p_1} \leq \omega \leq \omega_{p_2}, \\ H_{s_2}(\omega) = -\delta_{s_2} T_{S_2}(X_{s_2}(\omega)), & \omega_{s_2} \leq \omega \leq \pi. \end{cases} \tag{3.77}$$

Here, P, S_1 and S_2 are the orders of the Chebyshev polynomials in the passband, first stopband and second stopband, respectively. They are given by

$$P = \frac{N_p + 1}{2}, \tag{3.78}$$

$$S_1 = N_{s_1}, \tag{3.79}$$

$$S_2 = N_{s_2}, \tag{3.80}$$

where N_p, N_{s_1} and N_{s_2} represent the number of extrema in the frequency bands $\omega_{p_1} \leq \omega \leq \omega_{p_2}$, $0 \leq \omega \leq \omega_{s_1}$ and $\omega_{s_2} \leq \omega \leq \pi$ respectively. The functions, $X_p(\omega)$, $X_{s_1}(\omega)$ and $X_{s_2}(\omega)$, are needed to map the equiripple intervals of the

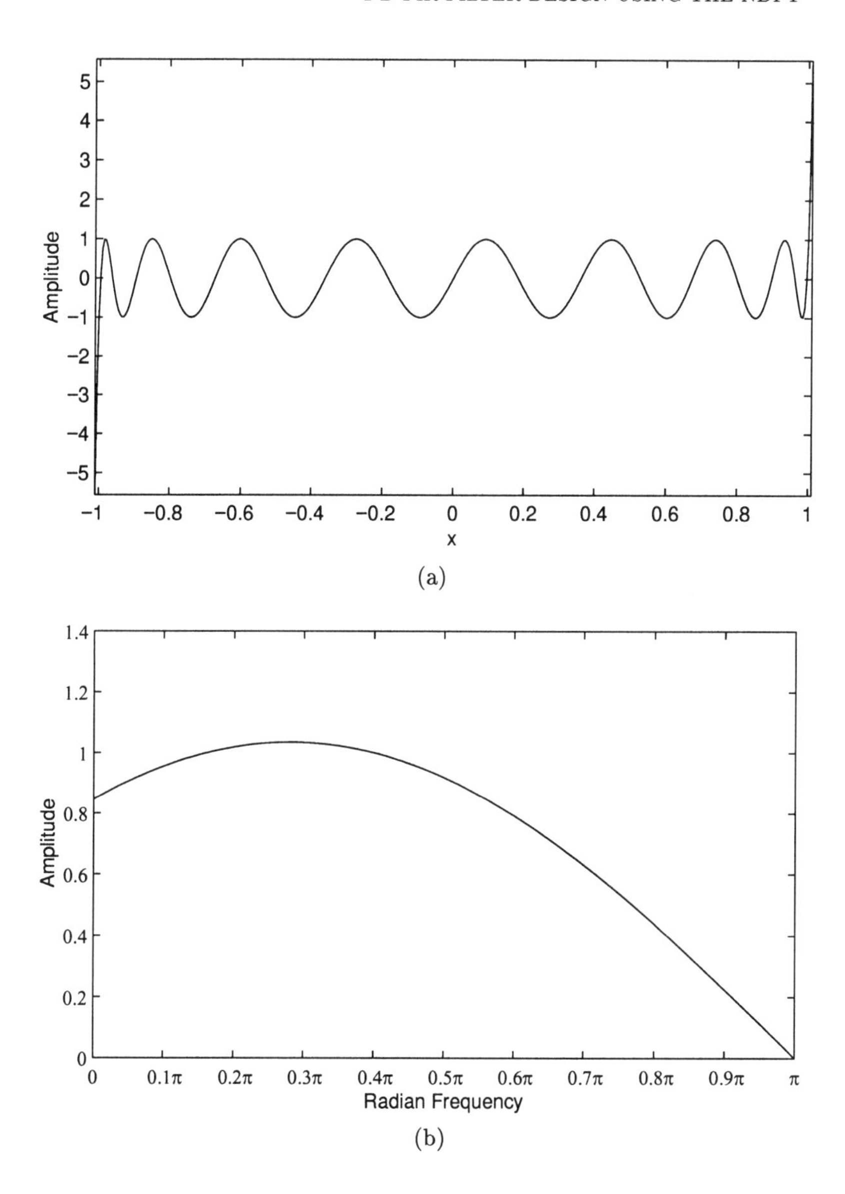

Figure 3.3. Type II lowpass filter design example. (a) Plot of Chebyshev polynomial $T_S(x)$ as a function of x. (b) Plot of $X_s(\omega)$ with normalized frequency $(\omega/2\pi)$.

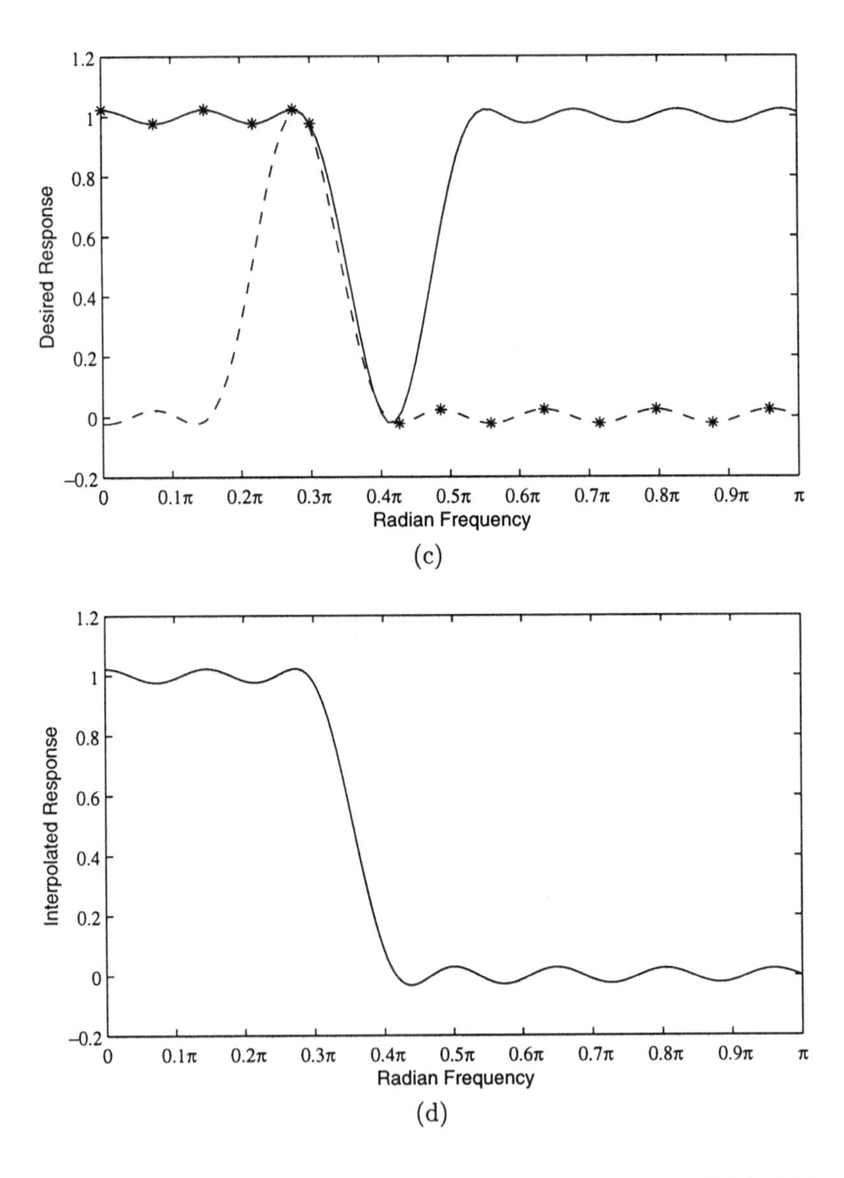

Figure 3.3. (continued) (c) Desired response approximated by $H_p(\omega)$ (solid line) in the passband, and $H_s(\omega)$ (dashed line) in the stopband. These functions are sampled at the locations denoted by "*". (d) Interpolated response of resulting lowpass filter.

Chebyshev polynomials to the respective frequency bands. They are chosen to be of the form

$$X_p(\omega) = A\cos(a\omega + b) + B, \tag{3.81}$$

$$X_{s_1}(\omega) = C\cos(c\omega + d) + D, \tag{3.82}$$

$$X_{s_2}(\omega) = E\cos(e\omega + f) + F, \tag{3.83}$$

where the twelve parameters, A, B, C, D, E, F, a, b, c, d, e, f, are obtained by imposing appropriate constraints on the three functions, $H_p(\omega)$, $H_{s_1}(\omega)$ and $H_{s_2}(\omega)$. This is outlined in the following twelve steps.

Step 1

$H_p(\omega)$ is symmetric with respect to the center of the passband,

$$\omega_m = \frac{1}{2}(\omega_{p_1} + \omega_{p_2}). \tag{3.84}$$

That is,

$$H_p(\omega_m + \omega) = H_p(\omega_m - \omega). \tag{3.85}$$

If we substitute for $H_p(\omega)$ from Eq. (3.77), we have

$$X_p(\omega_m + \omega) = X_p(\omega_m - \omega).$$

Substituting for $X_p(\omega)$ from Eq. (3.81), we obtain

$$\cos(a\omega_m + a\omega + b) = \cos(a\omega_m - a\omega + b),$$

or

$$b = \pi - a\omega_m. \tag{3.86}$$

Step 2

$H_{s_1}(\omega)$ is symmetric with respect to $\omega = 0$; that is,

$$H_{s_1}(\omega) = H_{s_1}(-\omega). \tag{3.87}$$

Using Eqs. (3.77) and (3.82), we obtain

$$d = \pi. \tag{3.88}$$

Step 3

$H_{s_2}(\omega)$ is symmetric with respect to $\omega = \pi$; that is

$$H_{s_2}(\pi + \omega) = H_{s_2}(\pi - \omega). \tag{3.89}$$

Using Eqs. (3.77) and (3.83), we obtain

$$f = \pi(1 - e). \tag{3.90}$$

Step 4

We constrain the value of $H_p(\omega)$ at $\omega = \omega_{p_1}$ so as to have an alternation at this band-edge:

$$H_p(\omega_{p_1}) = 1 - \delta_p. \tag{3.91}$$

Substituting for $H_p(\omega)$ from Eq. (3.77), we have

$$T_P(X_p(\omega_{p_1})) = 1,$$

or

$$X_p(\omega_{p_1}) = 1.$$

If we substitute for $X_p(\omega)$ from Eq. (3.81), we obtain

$$A\cos(a\omega_{p_1} + b) + B = 1.$$

Substituting for b from Eq. (3.86), we have

$$a = \frac{1}{(\omega_{p_1} - \omega_m)} \cos^{-1}\left(\frac{B-1}{A}\right). \tag{3.92}$$

Step 5

Similarly, we constrain the value of $H_{s_1}(\omega)$ at $\omega = \omega_{s_1}$ so that

$$H_{s_1}(\omega_{s_1}) = \delta_{s_1}. \tag{3.93}$$

Using Eqs. (3.77), (3.82) and (3.88), we obtain

$$c = \frac{1}{\omega_{s_1}} \cos^{-1}\left(\frac{D-1}{C}\right). \tag{3.94}$$

Step 6

We also constrain the value of $H_{s_2}(\omega)$ at $\omega = \omega_{s_2}$ so that

$$H_{s_2}(\omega_{s_2}) = \delta_{s_2}. \tag{3.95}$$

Using Eqs. (3.77), (3.83) and (3.90), we obtain

$$e = \frac{1}{(\omega_{s_2} - \pi)} \cos^{-1}\left(\frac{F-1}{E}\right). \tag{3.96}$$

Step 7

$H_p(\omega)$ must have an extremum at the center of the passband, $\omega = \omega_m$. Therefore,

$$H_p(\omega_m) = \begin{cases} 1 + \delta_p, & P = \text{odd}, \\ 1 - \delta_p, & P = \text{even}. \end{cases} \tag{3.97}$$

Using Eqs. (3.77), (3.81), and (3.86), we obtain

$$B = A - 1. \tag{3.98}$$

Step 8

We constrain $H_{s_1}(\omega)$ so that an extremum occurs at $\omega = 0$. Thus,

$$H_{s_1}(0) = \begin{cases} -\delta_{s_1}, & S_1 = \text{odd}, \\ \delta_{s_1}, & S_1 = \text{even}. \end{cases} \tag{3.99}$$

Using Eqs. (3.77), (3.82), and (3.88), we obtain

$$D = C - 1. \tag{3.100}$$

Step 9

Similarly, we constrain $H_{s_2}(\omega)$ so that an extremum occurs at $\omega = \pi$. That is,

$$H_{s_2}(\pi) = \begin{cases} -\delta_{s_2}, & S_2 = \text{odd}, \\ \delta_{s_2}, & S_2 = \text{even}. \end{cases} \tag{3.101}$$

Using Eqs. (3.77), (3.83), and (3.90), we obtain

$$F = E - 1. \tag{3.102}$$

We already have nine equations following from the nine conditions in Steps 1–9. In order to solve for the twelve parameters, we obtain three more equations by restricting the functions, $X_p(\omega)$, $X_{s_1}(\omega)$ and $X_{s_2}(\omega)$, to stay between $-\delta_s$ and $(1 + \delta_p)$ at all frequencies, where $\delta_s = \max(\delta_{s_1}, \delta_{s_2})$. This leads to the following three steps.

Step 10

$$\min(H_p(\omega)) = -\delta_s \tag{3.103}$$

Using Eqs. (3.77), (3.81) and (3.98), we obtain

$$A = \frac{1}{2}\left\{T_P{}^{-1}\left(\frac{1+\delta_s}{\delta_p}\right) + 1\right\}. \tag{3.104}$$

Step 11

$$\max(H_{s_1}(\omega)) = 1 + \delta_p \tag{3.105}$$

Using Eqs. (3.77), (3.82) and (3.100), we obtain

$$C = \frac{1}{2}\left\{ T_{S_1}{}^{-1}\left(\frac{1+\delta_p}{\delta_{s_1}}\right) + 1 \right\}. \tag{3.106}$$

Step 12

$$\max(H_{s_2}(\omega)) = 1 + \delta_p \tag{3.107}$$

Using Eqs. (3.77), (3.83) and (3.102), we obtain

$$E = \frac{1}{2}\left\{ T_{S_2}{}^{-1}\left(\frac{1+\delta_p}{\delta_{s_2}}\right) + 1 \right\}. \tag{3.108}$$

Thus, we have found closed-form expressions for all the twelve parameters. As in Section 3.3.1, N_p, N_{s_1} and N_{s_2} are found by weighting them proportionally to the sizes of the three respective frequency bands. The Chebyshev polynomial orders P, S_1 and S_2 are found from Eqs. (3.78), (3.79) and (3.80). The optimum filter response has a minimum of $(L+2)$ alternations, where $L = (N-1)/2$. This includes the alternations at the four band-edges ω_{s_1}, ω_{p_1}, ω_{p_2}, ω_{s_2}. Since the sum $N_p + N_{s_1} + N_{s_2}$ equals the total number of extrema in the range $0 \le \omega \le \pi$, it should be at least $(L-2) = (N-5)/2$. As with lowpass filter design, samples are placed at the extrema and band-edges in the range $0 \le \omega \le \pi$, so as to obtain $N_i = (N+1)/2$ linear equations. Note that the sample locations are given by closed-form expressions similar to those given in Section 3.3.1.

We can similarly design bandpass filters of Type II, by modifying this procedure, as done earlier for Type II lowpass filters. Moreover, this method can be extended easily to design bandstop filters and bandpass filters with multiple passbands, by mapping a separate Chebyshev polynomial to each frequency band, and then locating samples at the extrema and band-edges.

Example 3.4

Consider the design of a bandpass filter with the following specifications:

Filter length $N = 51$, Equal ripples $\delta_p = \delta_{s_1} = \delta_{s_2}$,
Lower stopband edge $\omega_{s_1} = 0.2\pi$, Lower passband edge $\omega_{p_1} = 0.25\pi$,
Upper passband edge $\omega_{p_2} = 0.6\pi$, Upper stopband edge $\omega_{s_2} = 0.65\pi$.

The orders of the three Chebyshev polynomials are: $P = 6$, $S_1 = 6$, $S_2 = 8$. Fig. 3.4(a) shows the three functions $H_p(\omega)$, $H_{s_1}(\omega)$ and $H_{s_2}(\omega)$, approximating the passband and two stopbands of the desired bandpass response. The frequency response of the resulting filter is shown in Fig. 3.4(b).

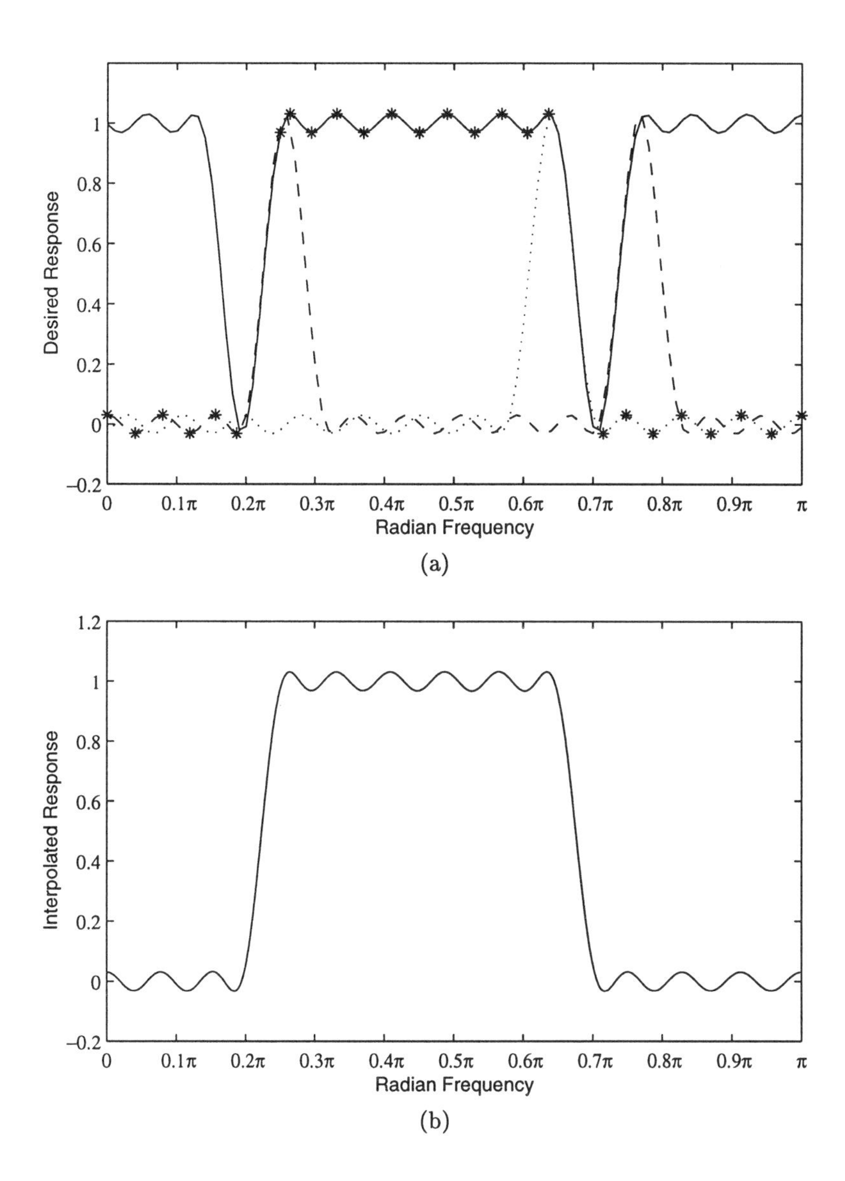

Figure 3.4. Bandpass filter design example. (a) Functions $H_p(\omega)$ (solid line), $H_{s_1}(\omega)$ (dashed line) and $H_{s_2}(\omega)$ (dotted line) approximate the three bands of the desired response. The sample locations are denoted by "$*$". (b) Interpolated response of resulting bandpass filter.

3.3.5 Third-band Filter Design

Mth-band filters are used for decimation and interpolation by a factor of M [Mintzer, 1982]. Half-band filter design ($M = 2$) was discussed in detail in Section 3.3.2. The impulse response $h[n]$ of an Mth-band filter has one out of M samples equal to zero. These filters are restricted to be of Type I. Let the filter have length N, passband edge ω_p, stopband edge ω_s, and peak ripples δ_p and δ_s in the passband and stopband, respectively. The main properties of an Mth-band filter are as follows:

$$h[n] = \begin{cases} 0, & n - \frac{N-1}{2} = \text{nonzero multiple of } M, \\ \frac{1}{M}, & n = \frac{N-1}{2}, \end{cases} \tag{3.109}$$

$$\omega_p + \omega_s = \frac{2\pi}{M}, \tag{3.110}$$

$$\delta_p \leq (M-1)\delta_s. \tag{3.111}$$

The frequency response has a passband $0 \leq \omega \leq \omega_p$ and $(M-1)$ stopbands interleaved with *don't care* bands. The stopbands are defined by

$$\left[-\omega_p + \frac{2k\pi}{M},\ \omega_p + \frac{2k\pi}{M}\right], \qquad k = 1, 2, \ldots, M-1.$$

As shown by Mintzer [Mintzer, 1982], Eq. (3.109) requires that

$$\sum_{k=0}^{M-1} H\left(\omega + \frac{2\pi k}{M}\right) = Mh[(N-1)/2]. \tag{3.112}$$

For a filter with a normalized passband,

$$h[(N-1)/2] = 1/M. \tag{3.113}$$

By taking advantage of the zero-valued coefficients, an Mth-band filter achieves a computational reduction of nearly $1/M$ when compared with a general FIR filter of the same length.

We can use the proposed nonuniform sampling method to design an Mth-band FIR filter by mapping separate Chebyshev polynomials to the passband and $(M-1)$ stopbands of the desired frequency response. Our technique can also utilize the knowledge of the zero-valued filter coefficients to obtain a reduced design time as compared with general FIR filter design. From the condition in Eq. (3.109), the number of independent filter coefficients in an Mth-band filter is given by

$$N_i = \frac{N-1}{2} - \left\lfloor \frac{N-1}{2M} \right\rfloor, \tag{3.114}$$

where $\lfloor x \rfloor$ denotes the integer closest to and smaller than x. Thus, we require N_i samples to design the filter. These samples are placed at the extrema and band-edges of the passband and stopbands; no samples are placed in the *don't care* bands, since we do not have any particular desired response for these bands.

For clarity, we now consider the case of third-band filter design, where $M = 3$. The frequency response has a passband $0 \leq \omega \leq \omega_p$, a stopband $(2\pi/3 - \omega_p) \leq \omega \leq (2\pi/3 + \omega_p)$ and a *don't care* band $(2\pi/3 + \omega_p) \leq \omega \leq \pi$. Thus, the stopband starts at $\omega_s = 2\pi/3 - \omega_p$.

Since this is a type I filter, its frequency response can be expressed as shown in Eq. (3.4). We express the real-valued amplitude response $A(\omega)$ in the form

$$A(\omega) = \begin{cases} H_p(\omega) = 1 - \delta_p T_P(X_p(\omega)), & 0 \leq \omega \leq \omega_p, \\ H_s(\omega) = -\delta_s T_S(X_s(\omega)), & \omega_s \leq \omega \leq \frac{2\pi}{3} + \omega_p. \end{cases} \tag{3.115}$$

The integers P and S are the orders of the Chebyshev polynomials in the passband and stopband, respectively. Since the widths of the passband and stopband are equal,

$$P = S. \tag{3.116}$$

By recognizing that there are M bands in the frequency response, and that P equals the number of extrema in the region $0 \leq \omega \leq \omega_p$, we obtain

$$P = \left\lfloor \frac{N-1}{2M} \right\rfloor, \tag{3.117}$$

where $M = 3$ for the third-band case. The functions $X_p(\omega)$ and $X_s(\omega)$ are used to map the equiripple intervals $-1 \leq x \leq 1$ of the Chebyshev polynomials $T_P(x)$ and $T_S(x)$ to the passband and stopband respectively:

$$X_p(\omega) = A\cos(a\omega + b) + B, \tag{3.118}$$

$$X_s(\omega) = C\cos(c\omega + d) + D. \tag{3.119}$$

The eight parameters, A, B, C, D, a, b, c, d, are obtained by imposing constraints on the functions, $H_p(\omega)$ and $H_s(\omega)$, so as to make them approximate the desired third-band response. This procedure is briefly outlined in the following eight steps.

Step 1

$H_p(\omega)$ is symmetric with respect to $\omega = 0$. Thus, we have

$$H_p(\omega) = H_p(-\omega). \tag{3.120}$$

Using Eqs. (3.115) and (3.118), we obtain

$$b = \pi. \tag{3.121}$$

Step 2

$H_s(\omega)$ is symmetric with respect to the center of the stopband $\omega = 2\pi/3$. Thus, we have

$$H_s\left(\frac{2\pi}{3} + \omega\right) = H_s\left(\frac{2\pi}{3} - \omega\right). \tag{3.122}$$

Using Eqs. (3.115), (3.119) and (3.121), we obtain

$$d = \pi\left(1 - \frac{2c}{3}\right). \tag{3.123}$$

Step 3

We constrain the value of $H_p(\omega)$ at $\omega = \omega_p$; that is,

$$H_p(\omega_p) = 1 - \delta_p. \tag{3.124}$$

From Eqs. (3.115), (3.118) and (3.121), we can simplify this to obtain

$$a = \frac{1}{\omega_p}\cos^{-1}\left(\frac{B-1}{A}\right). \tag{3.125}$$

Step 4

Similarly, we constrain the value of $H_s(\omega)$ at $\omega = \omega_s$, so that

$$H_s(\omega_s) = \delta_s. \tag{3.126}$$

Using Eqs. (3.115) and (3.119), we have

$$C\cos(c\omega_s + d) + D = 1.$$

On applying Eqs. (3.123) and (3.110), we obtain

$$c = -\frac{1}{\omega_p}\cos^{-1}\left(\frac{D-1}{C}\right). \tag{3.127}$$

Step 5

We constrain $H_p(\omega)$ to have an extremum at $\omega = 0$. Thus, we have

$$H_p(0) = \begin{cases} 1 + \delta_p, & P = \text{odd}, \\ 1 - \delta_p, & P = \text{even}. \end{cases} \tag{3.128}$$

Using Eqs. (3.115) and (3.118) and (3.121), we have

$$B = A - 1. \tag{3.129}$$

Step 6

Similarly, we constrain $H_s(\omega)$ so that an extremum occurs at the center of the stopband, $\omega = 2\pi/3$. Therefore,

$$H_s(2\pi/3) = \begin{cases} -\delta_s, & S = \text{odd}, \\ \delta_s, & S = \text{even}. \end{cases} \tag{3.130}$$

Using Eqs. (3.115) and (3.119), and (3.123), we obtain

$$D = C - 1. \tag{3.131}$$

Two more equations are obtained by requiring that $H_p(\omega)$ and $H_s(\omega)$ stay between $-\delta_s$ and $(1 + \delta_p)$ at all frequencies.

Step 7

$$\min(H_p(\omega)) = -\delta_s \tag{3.132}$$

Using Eqs. (3.115), (3.118) and (3.129), we simplify this to obtain

$$A = \frac{1}{2}\left\{{T_P}^{-1}\left(\frac{1+\delta_s}{\delta_p}\right) + 1\right\}, \tag{3.133}$$

Step 8

$$\max(H_s(\omega)) = 1 + \delta_p \tag{3.134}$$

Using Eqs. (3.115), (3.119) and (3.131), we obtain

$$C = \frac{1}{2}\left\{{T_S}^{-1}\left(\frac{1+\delta_p}{\delta_s}\right) + 1\right\}. \tag{3.135}$$

Thus, the eight parameters have been expressed in closed form. While choosing the sample locations, we must carefully consider the relation in Eq. (3.112) to pick the independent region of the frequency response. For the third-band case, we place samples at the extrema in the regions $0 \leq \omega \leq \omega_p$ and $\omega_s \leq \omega < 2\pi/3$, since the response in the rest of the stopband is fully determined by the response in these regions. We compute the sample locations using closed-form expressions, as discussed earlier.

Example 3.5

Consider the design of a third-band filter with the following specifications:

Filter length $N = 29$, Passband edge $\omega_p = 0.27\pi$

Stopband from $\omega_{s_1} = 0.3967\pi$ to $\omega_{s_2} = 0.9367\pi$.

The orders of the Chebyshev polynomials for the passband and stopband are: $P = S = 5$. Fig. 3.5(a) shows the functions, $H_p(\omega)$ and $H_s(\omega)$, which approximate the passband and stopband of the desired third-band response. We solve for the ten independent filter coefficients by placing ten samples as shown in this figure. Fig. 3.5(b) shows that the resulting filter has a nearly equiripple response in the passband and stopband. Note the *don't care* band in the frequency response.

3.4 RESULTS

In many cases, the filters designed by the proposed nonuniform frequency sampling method are *very close to optimal* equiripple filters. This was illustrated in Examples 3.1–3.5. The design time required by our method is *much lower*, when compared to iterative optimization methods such as the Parks-McClellan algorithm [McClellan et al., 1973]. Fig. 3.6 shows the variation of design time with filter length for the proposed method and Parks-McClellan algorithm. This plot was obtained by recording the times taken to design a lowpass filter with fixed band edges and increasing length. The simulations were performed on a SPARC-2 workstation. From the comparison presented in Fig. 3.6, it is clear that the proposed method is particularly useful for designing long filters, since iterative routines need excessively large amounts of time in such cases. Our method can be used to design nearly optimal filters in a much shorter time.

All symmetries in the filter impulse response are utilized in the proposed method, so that we need to solve for only the *independent filter coefficients*. For example, the presence of alternating zeros in the impulse response of half-band filters has been utilized, leading to a reduction in the design time by a factor of half. This also demonstrates the *flexibility* of nonuniform frequency sampling. In third-band filter design, we chose not to place any samples in the *don't care* frequency band.

In the following three subsections, we first present some filter design examples and compare the filters designed by the proposed NDFT method with other existing designs. Then we show that the proposed choice of extremal frequencies can serve as a good starting point for the Parks-McClellan algorithm. Finally, we present justification for the choice of extrema as sampling points in our filter design method.

3.4.1 Comparison With Other Design Methods

We begin by comparing the NDFT-based design method with other frequency sampling designs. A lowpass filter with specifications as given in Example 3.1 is designed using three uniform frequency sampling methods mentioned in Section 3.2:

(1) Original uniform frequency sampling with sample values of unity in the passband and zeros in the stopband.

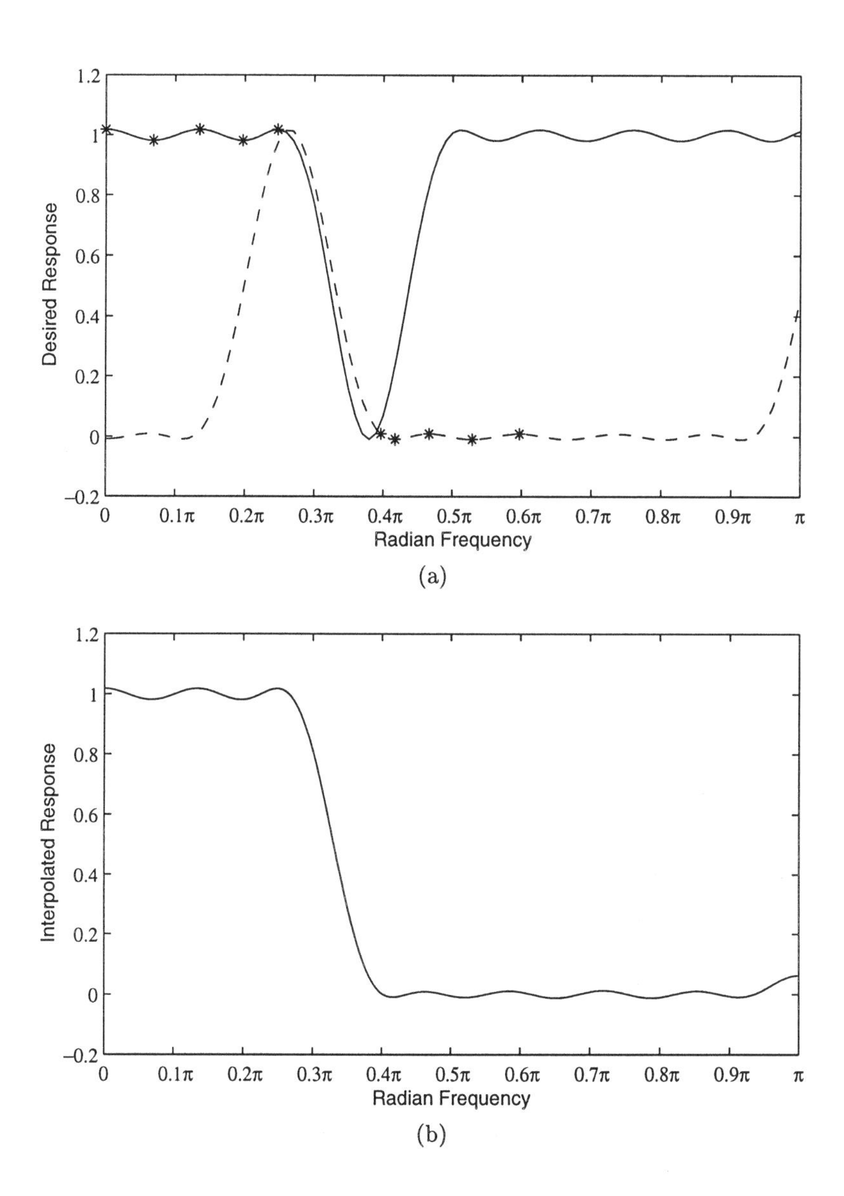

Figure 3.5. Third-band filter design example. (a) Functions $H_p(\omega)$ (solid line) and $H_s(\omega)$ (dashed line) approximate the passband and stopband of the desired response. The sample locations are denoted by "$*$". (b) Interpolated response of resulting third-band filter.

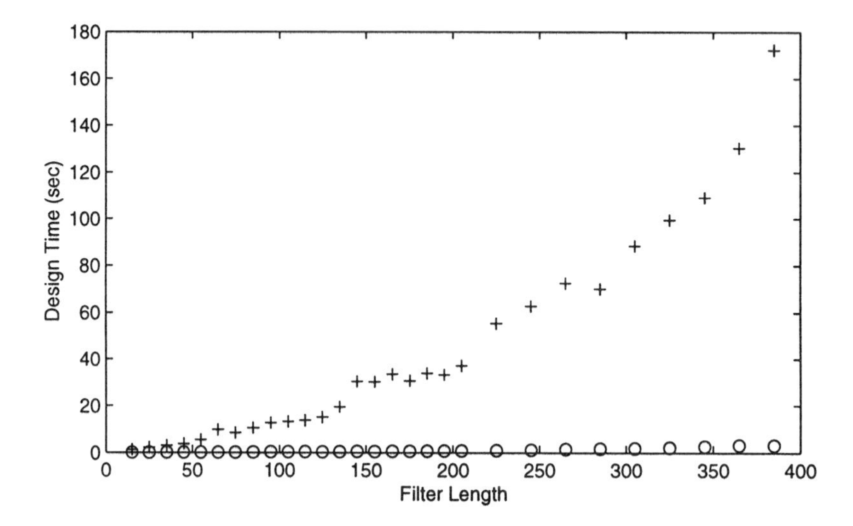

Figure 3.6. Plot of filter design time with filter length, for the NDFT method ("o") and Parks-McClellan algorithm ("+").

(2) Uniform frequency sampling using linear programming to optimize the values of the samples in the transition band [Rabiner et al., 1970].

(3) Modified uniform frequency sampling using analytic functions to approximate the desired frequency response [Lightstone et al., 1994].

For Method 2, there is only one transition sample with the given specifications. Its value is chosen to be 0.39, so as to minimize the peak stopband ripple. In Method 3, the analytic functions generated in the first step of the NDFT-based design method are sampled uniformly. Since some of these samples are in the transition band, an additional function is required to approximate the frequency response in this band. This function is chosen to be a weighted linear combination of the passband and stopband functions [Lightstone et al., 1994].

The frequency responses of the resulting filters are shown in Figs. 3.7—3.9. The filter designed by Method 1 has large ripples near the band edges. They are decreased by introducing the transition sample in Method 2. Method 3 provides better control over the band edges. Fig. 3.10 shows the filter designed by the NDFT method. Table 3.1 provides a comparison of the band edges and attenuation actually attained by these filters. A_p is the passband attenuation, defined as

$$A_p = -\log_{10}(1 - \delta_p), \tag{3.136}$$

where δ_p is the maximum ripple in the passband. A_s is the stopband attenuation, defined as

$$A_s = -\log_{10} \delta_s, \tag{3.137}$$

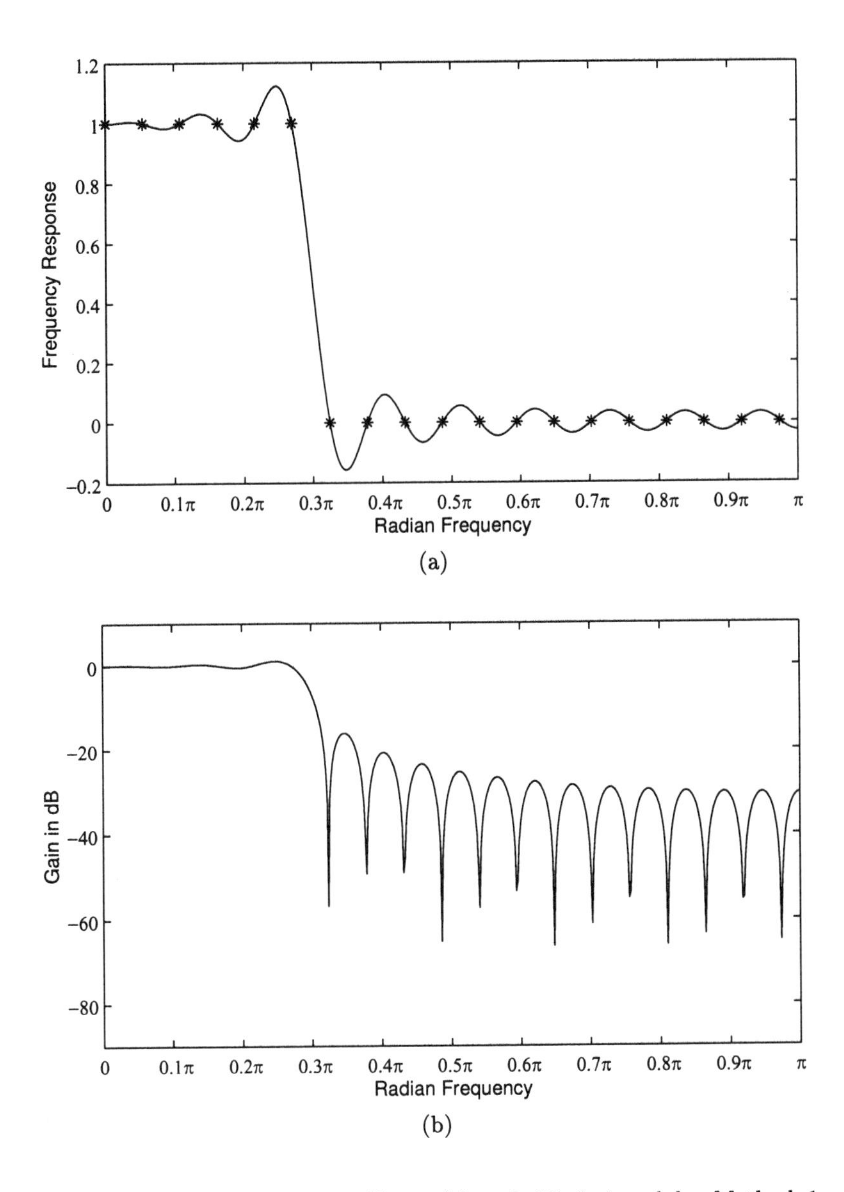

Figure 3.7. Type I lowpass filter of length 37 designed by Method 1. (a) Frequency response, with samples denoted by "*". (b) Gain response.

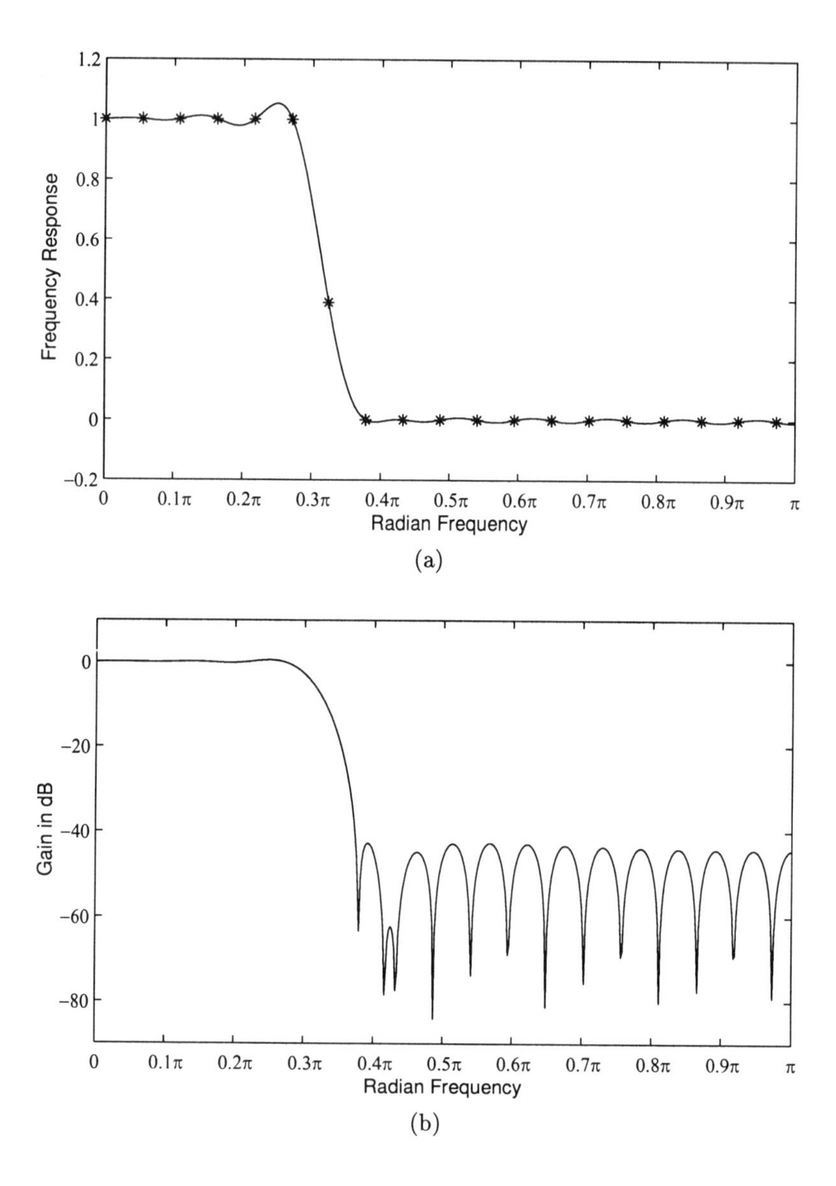

Figure 3.8. Type I lowpass filter of length 37 designed by Method 2. (a) Frequency response, with samples denoted by "∗". (b) Gain response.

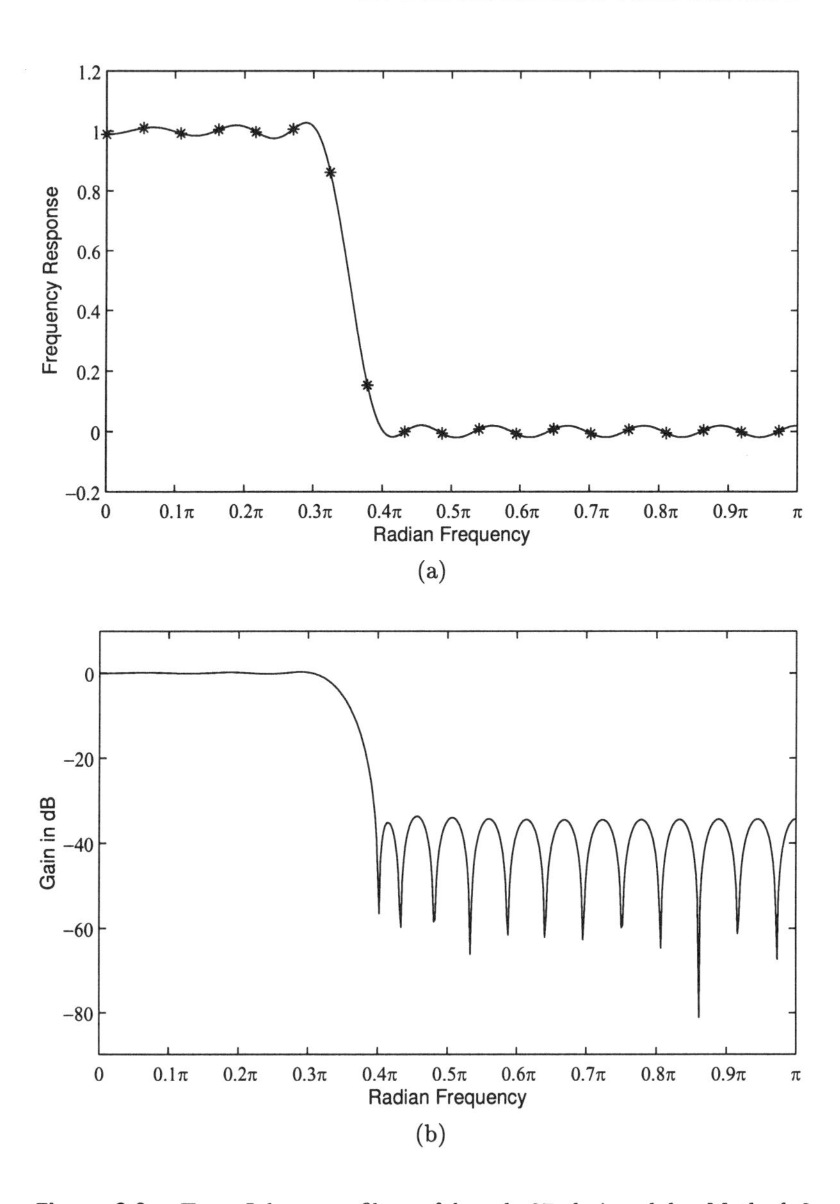

Figure 3.9. Type I lowpass filter of length 37 designed by Method 3. (a) Frequency response, with samples denoted by "∗". (b) Gain response.

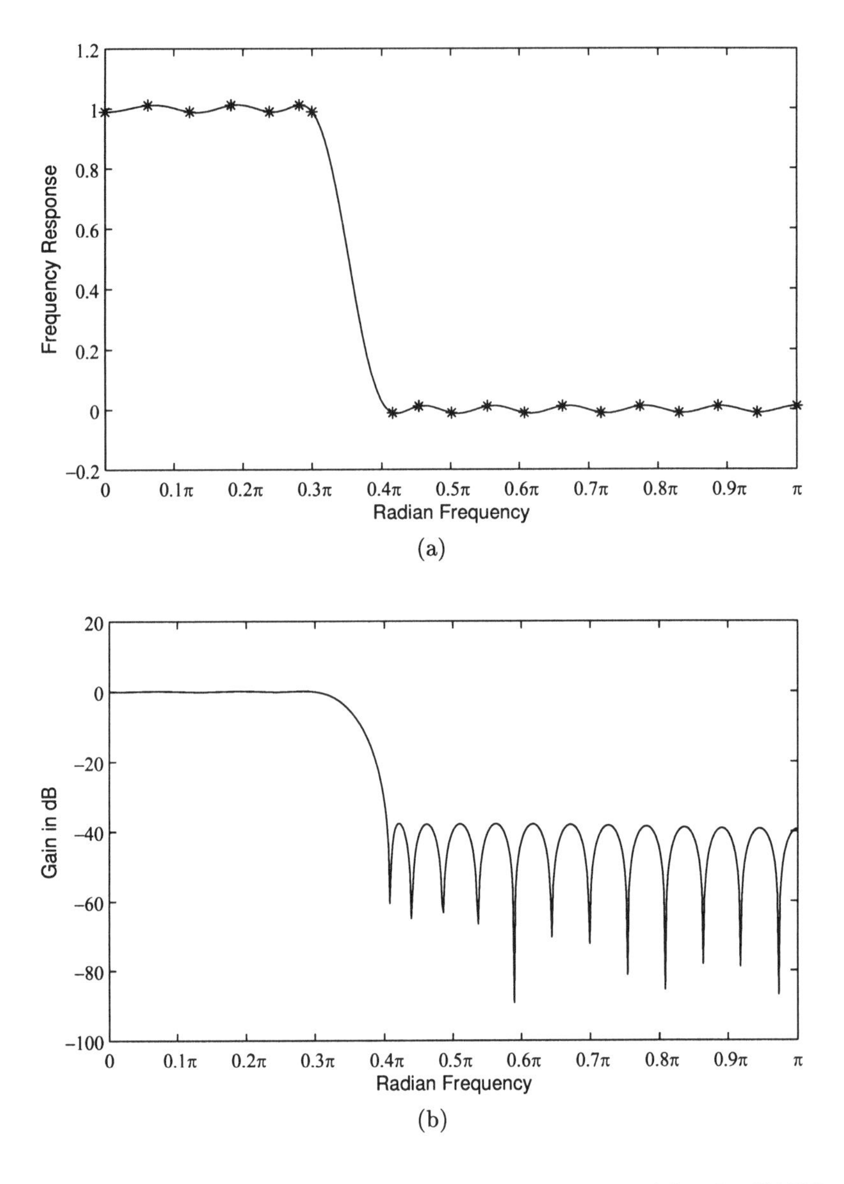

Figure 3.10. Type I lowpass filter of length 37 designed by the NDFT method. (a) Frequency response, with samples denoted by "$*$". (b) Gain response.

Table 3.1. Performance comparison for frequency sampling designs of Type I lowpass filter.

Method	ω_p (π)	ω_s (π)	A_p (dB)	A_s (dB)
Method 1	0.2790	0.3374	1.1512	16.0977
Method 2	0.2782	0.3740	0.4671	42.6863
Method 3	0.3103	0.3963	0.2404	33.6428
NDFT	0.3006	0.4039	0.1121	37.7473

where δ_s is the maximum ripple in the stopband. Clearly, the NDFT method provides the lowest A_p, the highest A_s, and band edges that are closest to the desired specifications. Methods 1 and 2 provide inadequate control over the band edges and no control over the ratio of the passband and stopband ripples. Therefore, we do not include them in the remaining comparisons.

To further demonstrate the performance of the NDFT-based nonuniform frequency sampling method, we consider design examples for each type of filter discussed in Section 3.3. Three filter design methods are used—the NDFT method, Parks-McClellan algorithm (PM) and modified uniform frequency sampling (DFT) [Lightstone et al., 1994]. The examples are shown on pages 86–90. We compare the performances of the three methods in Tables 3.2–3.6. The design times required are shown in the last column of each table. As expected, the design times are the largest when iterative Parks-McClellan algorithm is used. Much smaller times are required by the frequency sampling methods, both uniform and nonuniform. The filters designed by nonuniform frequency sampling come much closer to optimal filters. Note that the stopband attenuations A_s of the filters designed by the NDFT and PM methods differ by only about 1 dB. Figs. 3.11–3.15 show the gains of the filters designed by the NDFT method. The flexibility of nonuniform sampling is illustrated well in third-band filter design.

Example 3.6 Type I lowpass filter design

Desired specifications:
Filter length $N = 77$, Ripple ratio $\delta_p/\delta_s = 2$,
Passband edge $\omega_p = 0.4\pi$, Stopband edge $\omega_s = 0.45\pi$.

Table 3.2. Performance comparison for Type I lowpass filter design.

Method	ω_p (π)	ω_s (π)	A_p (dB)	A_s (dB)	Time (sec)
NDFT	0.4001	0.4520	0.1200	42.0572	0.2833
PM	0.4000	0.4501	0.1416	41.8731	8.5833
DFT	0.3990	0.4467	0.1138	34.5563	0.1167

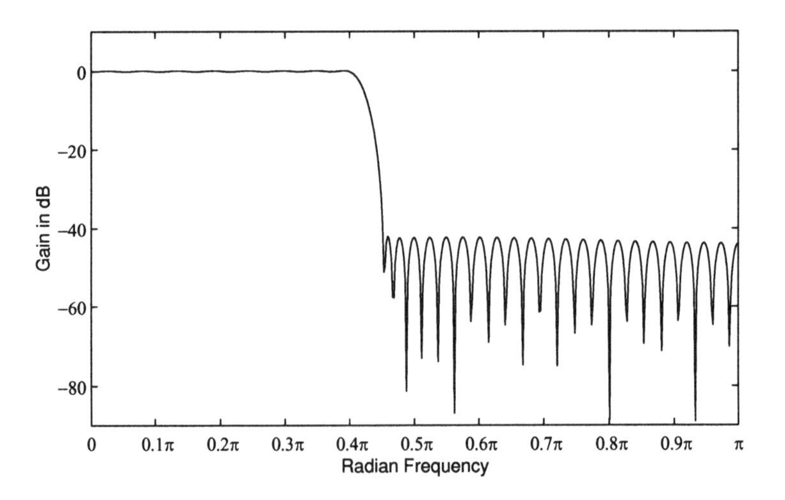

Figure 3.11. Gain of Type I lowpass filter of length 77 designed by the NDFT method.

Example 3.7 Half-band lowpass filter design

Desired specifications:
Filter length $N = 55$,
Passband edge $\omega_p = 0.46\pi$, Stopband edge $\omega_s = 0.54\pi$.

Table 3.3. Performance comparison for half-band lowpass filter design.

Method	ω_p (π)	ω_s (π)	A_p (dB)	A_s (dB)	Time (sec)
NDFT	0.4593	0.5407	0.0544	44.0894	0.0833
PM	0.4600	0.5400	0.0572	43.6580	11.4833
DFT	0.4619	0.5381	0.1052	38.3880	0.1000

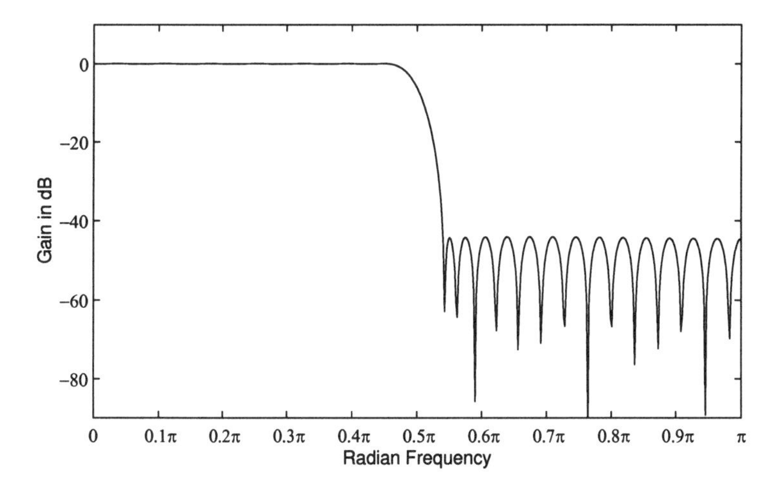

Figure 3.12. Gain of half-band lowpass filter of length 55 designed by the NDFT method.

Example 3.8 Type II lowpass filter design

Desired specifications:
Filter length $N = 46$, Ripple ratio $\delta_p/\delta_s = 3$,
Passband edge $\omega_p = 0.35\pi$, Stopband edge $\omega_s = 0.45\pi$.

Table 3.4. Performance comparison for Type II lowpass filter design.

Method	ω_p (π)	ω_s (π)	A_p (dB)	A_s (dB)	Time (sec)
NDFT	0.3501	0.4531	0.0799	49.0769	0.1833
PM	0.3500	0.4501	0.0906	49.2479	6.3667
DFT	0.3512	0.4467	0.1288	45.0239	0.1167

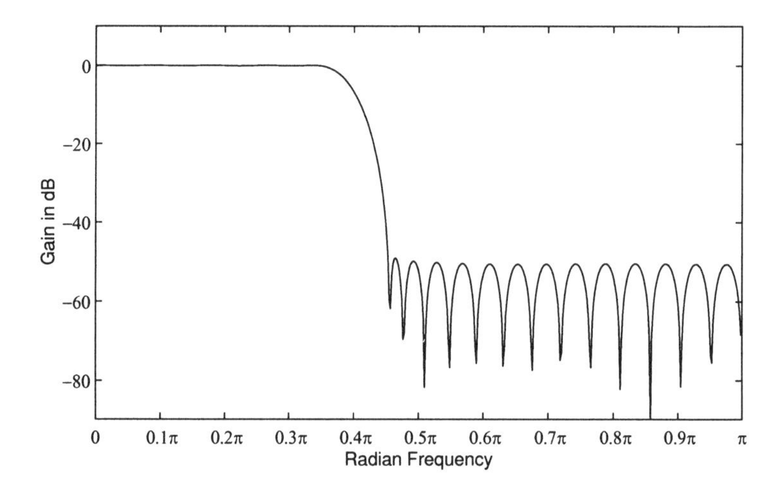

Figure 3.13. Gain of Type II lowpass filter of length 46 designed by the NDFT method.

Example 3.9 Bandpass filter design

Desired specifications:
Filter length $N = 63$, Equal ripples $\delta_p = \delta_{s_1} = \delta_{s_2}$,
Lower stopband edge $\omega_{s_1} = 0.25\pi$, Lower passband edge $\omega_{p_1} = 0.3\pi$,
Upper passband edge $\omega_{p_2} = 0.65\pi$, Upper stopband edge $\omega_{s_2} = 0.7\pi$.

Table 3.5. Performance comparison for bandpass filter design.

Method	ω_{s1} (π)	ω_{p1} (π)	ω_{p2} (π)	ω_{s2} (π)	A_p (dB)	A_{s1} (dB)	A_{s2} (dB)	Time (sec)
NDFT	0.2507	0.3000	0.6500	0.6999	0.2267	31.6143	31.5478	0.1667
PM	0.2500	0.3000	0.6500	0.7000	0.2205	32.0339	31.9934	5.9833
DFT	0.2503	0.3002	0.6510	0.6995	0.2314	31.1514	30.0499	0.1500

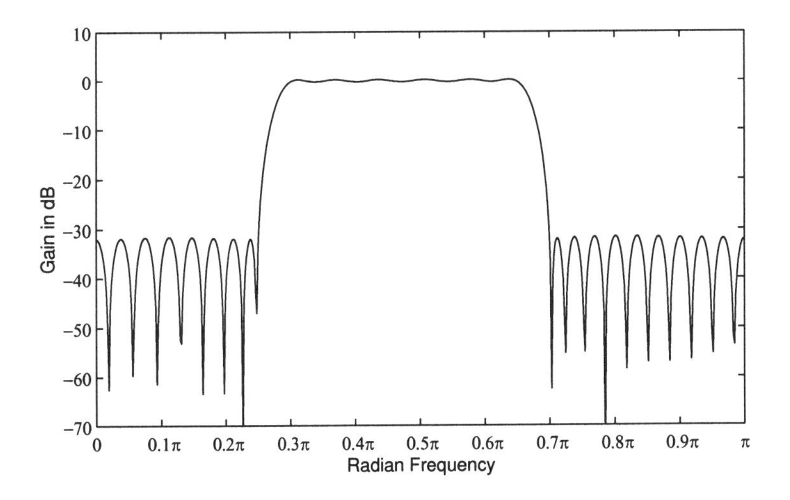

Figure 3.14. Gain of bandpass filter of length 63 designed by the NDFT method.

Example 3.10 Third-band filter design

Desired specifications:
Filter length $N = 23$, Passband edge $\omega_p = 0.2\pi$
Stopband from $\omega_{s_1} = 0.4667\pi$ to $\omega_{s_2} = 0.8667\pi$.

Table 3.6. Performance comparison for third-band filter design.

Method	ω_p (π)	ω_{s1} (π)	ω_{s2} (π)	A_p (dB)	A_s (dB)	Time (sec)
NDFT	0.2014	0.4662	0.8709	0.0139	60.8634	0.1333
PM	0.2000	0.4663	0.8673	0.0139	60.9417	3.1500

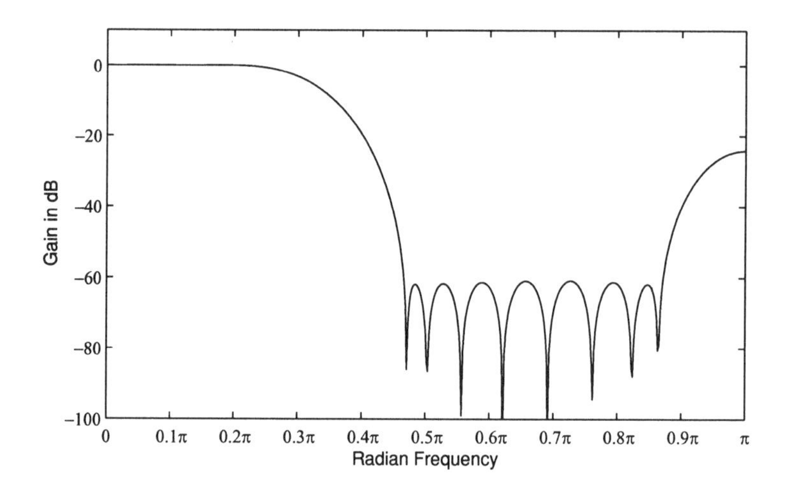

Figure 3.15. Gain of third-band filter of length 23 designed by the NDFT method.

3.4.2 A Good Starting Point for the Parks-McClellan Algorithm

In the NDFT-based filter design method, we have proposed a choice of extremal frequencies based on a nonuniform sampling of analytic functions derived from Chebyshev polynomials. This set of extrema can also be used as a good starting point for the Parks-McClellan algorithm, if exactly optimal filters are to be designed. We have done simulations to compare the performances of this algorithm with two different starting points: (a) the proposed nonuniform choice of extrema, and (b) the standard uniform choice of extrema. Our results show that Choice (a) *decreases* the number of iterations required, as well as the design time. Fig. 3.16 shows the variation of each of these quantities with filter length using the two starting points. These plots were obtained by designing a lowpass filter with fixed band edges, and gradually increasing filter length.

3.4.3 Choice of Extrema as Sample Locations

The results shown in this section demonstrate that the filters designed by our nonuniform frequency sampling method are very close to optimal minimax filters. We now consider the reasons behind our choice of extrema as sample locations. Note that the Parks-McClellan algorithm is based on the fact that an optimum minimax filter must have alternations at the extremal frequencies [Oppenheim and Schafer, 1989]. It starts with an initial guess for the extrema, fits a polynomial, checks the resulting ripple δ, and then searches for the new extremal frequencies. This cycle is repeated until δ does not change from its previous value by more than a specified small amount. Thus, the initial guess of extrema is improved in successive iterations until the optimal filter is obtained.

In our method, we generate a good approximation to the desired response, and then sample this at the extrema to obtain the linear equations required to solve for the filter coefficients. We now show that this choice of sample locations *minimizes* the sensitivity of the frequency response to a perturbation in the sample locations. Let $D(z)$ be the desired response, and $H(z)$ be the actual filter response obtained, i.e., the resulting filter polynomial of order $N-1$. We solve for the N coefficients of $H(z)$ by taking N samples:

$$H(z_k) = D(z_k), \qquad k = 0, 1, \ldots, N-1. \tag{3.138}$$

Thus, the function $H(z)$ can be written explicitly as a function of not only the variable z, but also the sample locations $z_0, z_1, \ldots, z_{N-1}$. This gives the following form:

$$H(z) = H(z_0, z_1, \ldots, z_{N-1}, z). \tag{3.139}$$

Therefore, the samples can be expressed as

$$H(z_k) = H(z_0, z_1, \ldots, z_{N-1}, z_k). \tag{3.140}$$

Now, we differentiate both sides of Eq. (3.138) with respect to z_k to get

$$\frac{\partial H(z_k)}{\partial z_k} = \frac{\partial D(z_k)}{\partial z_k} = \left.\frac{\partial D(z)}{\partial z}\right|_{z=z_k}. \tag{3.141}$$

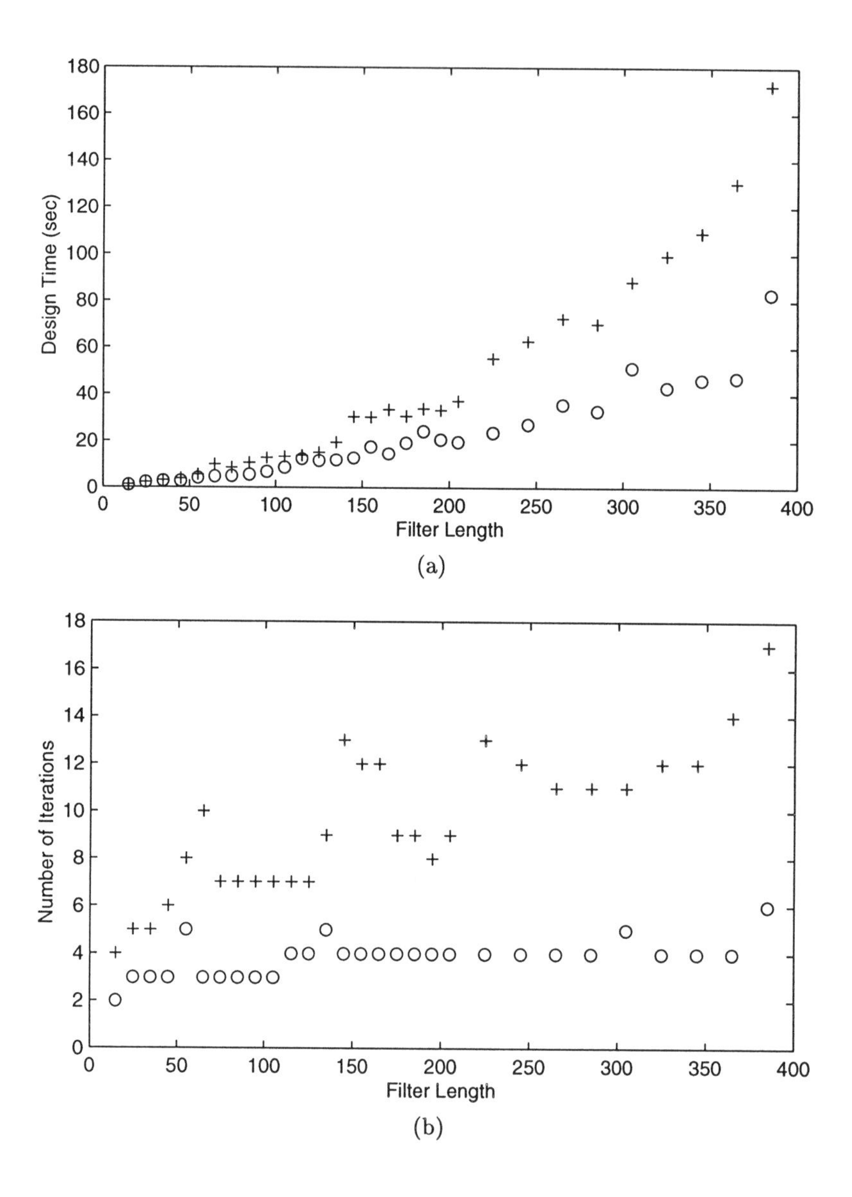

Figure 3.16. Comparison of different starting points for the Parks-McClellan algorithm — "o" with proposed nonuniformly spaced extrema, and "+" with standard uniformly spaced extrema. (a) Plot of design time with filter length. (b) Plot of number of iterations with filter length.

Differentiating both sides of Eq. (3.140) with respect to z_k, we get

$$\frac{\partial H(z_k)}{\partial z_k} = \left.\frac{\partial H(z)}{\partial z_k}\right|_{z=z_k} + \left.\frac{\partial H(z)}{\partial z}\right|_{z=z_k}. \tag{3.142}$$

Comparing Eqs. (3.141) and (3.142), we obtain

$$\left.\frac{\partial D(z)}{\partial z}\right|_{z=z_k} = \left.\frac{\partial H(z)}{\partial z_k}\right|_{z=z_k} + \left.\frac{\partial H(z)}{\partial z}\right|_{z=z_k}, \tag{3.143}$$

or

$$\left.\frac{\partial H(z)}{\partial z_k}\right|_{z=z_k} = \left.\frac{\partial D(z)}{\partial z}\right|_{z=z_k} - \left.\frac{\partial H(z)}{\partial z}\right|_{z=z_k} = \left.\frac{\partial E(z)}{\partial z}\right|_{z=z_k}, \tag{3.144}$$

where $E(z)$ is the interpolation error function defined as

$$E(z) = D(z) - H(z). \tag{3.145}$$

The quantity on the left hand side of Eq. (3.144) is the sensitivity of the frequency response at the point z_k to a small perturbation in the location of the k-th sample. This sensitivity equals the difference in slopes of $D(z)$ and $H(z)$ at $z = z_k$, i.e., the slope of the interpolation error at this point. If we sample at the extrema of $D(z)$, the slope of $D(z)$ is zero at these points. Since these points lie close to the extrema of $H(z)$ actually obtained, the slope of $H(z)$ is also nearly zero there. Thus the resulting sensitivities at these points are close to zero. Therefore the extrema are good locations for the samples from the sensitivity point of view.

Finally, we present a numerical example to demonstrate that if we perturb the location of a sample away from the extremum, the peak ripple in that particular frequency band increases.

Example 3.11

Consider the design of a lowpass filter with the following specifications:

Filter length $N = 37$, Ripple ratio $\delta_p/\delta_s = 1$,
Passband edge $\omega_p = 0.4\pi$, Stopband edge $\omega_s = 0.5\pi$.

By using the NDFT-based design method, we obtain a nearly equiripple filter with the following values for the peak attenuation in the passband and stopband, respectively:

$A_p = 0.0956$ dB, $A_s = 36.9301$ dB.

Now, we vary the location of one sample in the passband on either side of the extremum, and redesign the filter for each new sample location. Fig. 3.17(a) shows the variation of A_p with change in sample frequency about the extremum. This plot shows that the lowest passband attenuation is obtained when the sample is at the extremum. This process is repeated by varying the location of one sample in the stopband about the extremum. Fig. 3.17(b) shows that the resulting stopband attenuation is maximum when the sample is at the extremal frequency.

Next, we simultaneously vary the locations of two adjacent samples in the passband about their respective extremal positions. The filter is redesigned for each new pair of locations. Fig. 3.18(a) shows the values of A_p obtained in these situations. Again, the passband attenuation is minimum when both samples are at the extrema. This procedure is repeated by varying simultaneously, the locations of two adjacent samples in the stopband. Fig. 3.18(b) shows the resulting stopband attenuation, which reaches a maximum when both samples are at the extrema.

This example demonstrates that the extrema form a good choice of sample locations for filter design, from the point of view of *minimizing* the peak ripple.

3.5 SUMMARY

In this chapter, we utilized the concept of the NDFT to develop a new nonuniform frequency sampling method for designing 1-D FIR filters. Although there have been some earlier efforts to design 1-D filters using nonuniform frequency samples [Rabiner et al., 1970], no working design technique existed before. The proposed design method produces nearly optimal filters, as demonstrated by the design examples presented in Section 3.4.1. Since the design time required is much smaller than for optimal (minimax) filters, this method is particularly useful for designing long filters. Moreover, the proposed choice of nonuniform extremal frequencies can be used as a good starting point for the Parks-McClellan algorithm, if truly optimal filters are desired. The results obtained in this chapter also illustrate the advantages of nonuniform frequency sampling over uniform sampling. In Chapter 4, we shall see how the flexibility available in locating the frequency samples becomes even more important, when applied to the 2-D case.

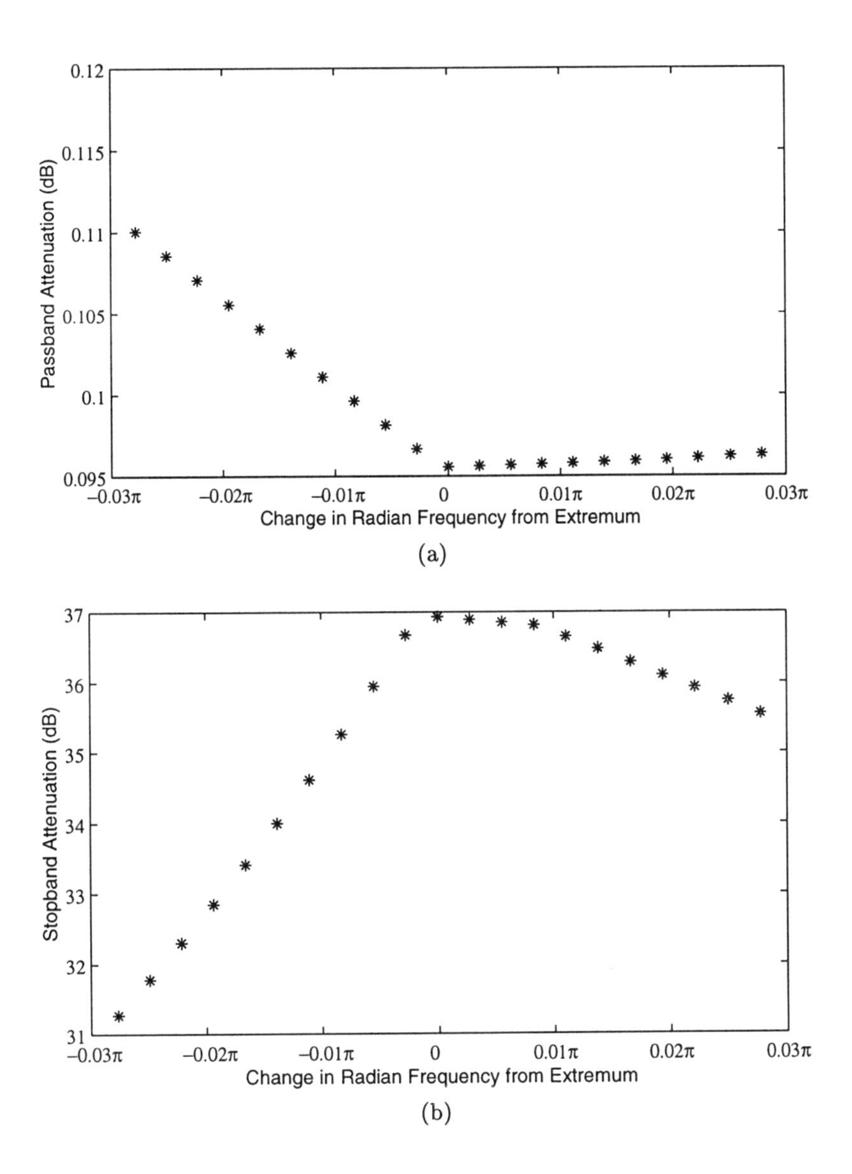

Figure 3.17. Effects of change of location of one sample away from the extremum. (a) Variation of passband attenuation A_p, with change in frequency of a sample in the passband. (b) Variation of stopband attenuation A_s, with change in frequency of a sample in the stopband.

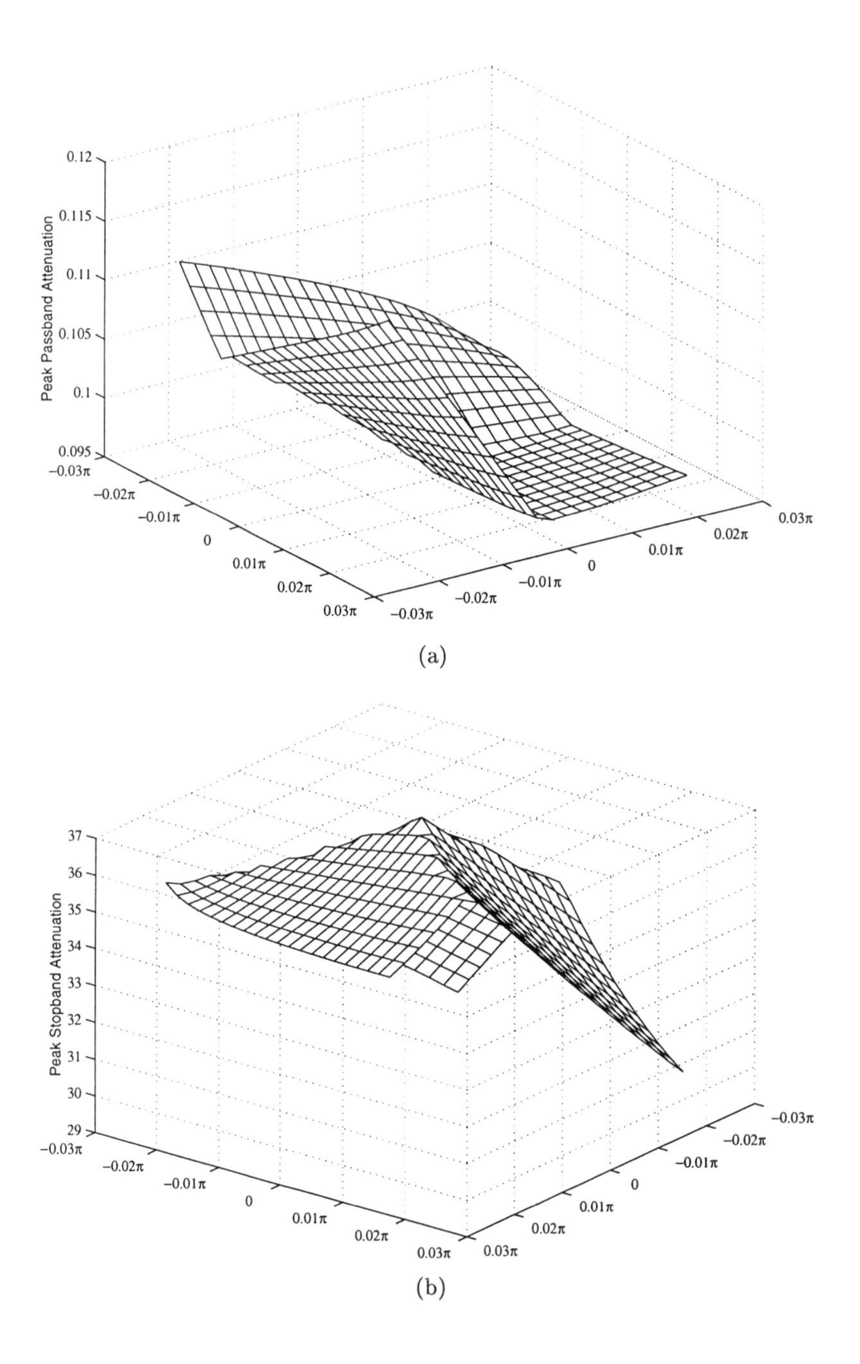

Figure 3.18. Effects of change of location of two adjacent samples away from the extrema. (a) Variation of passband attenuation A_p, with changes in frequency of two samples in the passband. (b) Variation of stopband attenuation A_s, with changes in frequency of two samples in the stopband.

4 2-D FIR FILTER DESIGN USING THE NDFT

4.1 INTRODUCTION

Two-dimensional (2-D) digital filters find applications in diverse areas such as image processing and coding, robotics and computer vision, seismology, sonar, radar and astronomy. In applications such as image processing, preservation of the phase information is important [Huang et al., 1975]. In this respect, FIR filters offer a distinct advantage over IIR filters, since they can easily attain linear-phase or zero-phase characteristics. Besides, they are guaranteed to be stable. In this chapter, we use the 2-D NDFT, defined in Chapter 2 (Section 2.5), to design 2-D FIR filters by nonuniform frequency sampling of the specified frequency response.

There exist four standard approaches for designing 2-D FIR filters—windowing, frequency sampling, frequency transformation, and optimal filter design methods. In the window method, a 2-D window is typically obtained from a 1-D window by separable or nonseparable techniques. The infinite-extent impulse response of an ideal 2-D filter is then multiplied by this window to obtain a 2-D FIR filter. Although straightforward, this method suffers from a lack of control over the frequency domain specifications. Besides, it can only design filters with separable or circularly symmetric frequency responses. Frequency sampling is a conceptually simple approach to 2-D FIR filter design. It includes a variety of existing methods that use uniform as well as nonuniform frequency sampling. We shall take a close look at these methods in Section 4.2. Filter design by frequency transformation does not have a 1-D counterpart. This is an attractive, practical method of designing a 2-D filter by applying a frequency transformation function to a 1-D filter. The function controls the shape of the contours of the 2-D frequency response. The amplitude characteristics of the 1-D filter are preserved in the 2-D filter. These three methods are not optimal. However, in contrast to the case of 1-D filters, development of a practical, reliable algorithm to design optimal 2-D filters still remains an area for active research. The problem of designing an optimal 2-D filter is much more complex than in the 1-D case. The powerful alternation theorem does not apply in 2-D, and the minimax solution is not unique. Consequently, existing iterative algorithms of the Remez exchange type are very intensive computationally, and do not always converge to a correct solution.

Earlier efforts in nonuniform 2-D frequency sampling design have involved either constrained sampling structures which reduce computational complexity [Zakhor and Alvstad, 1992; Angelides, 1994; Rozwood et al., 1991; Diamessis et al., 1987], or a linear least squares approach that guarantees unique interpolation [Zakhor and Alvstad, 1992]. Our approach involves generalized frequency sampling, where the samples are placed on contour lines that match the desired shape of the passband or stopband of the 2-D filter. The proposed method produces nonseparable 2-D filters with *good passband shapes* and *low peak ripples*. Filters of good quality are obtained, even for small support sizes. This is important since such filters are most likely to be used in practical filtering applications.

This chapter is organized as follows. We discuss existing methods for frequency sampling design of 2-D FIR filters in Section 4.2. The general strategy for the proposed 2-D nonuniform sampling method is outlined in Section 4.3. Details of the procedure for designing 2-D filters with square, circular, diamond, fan, and elliptically-shaped passbands are given in Sections 4.4– 4.7. We demonstrate the potential of our method by presenting filter design examples and comparisons with filters obtained by other design methods. In Section 4.8, we outline applications of 2-D filters. The performances of the square and diamond filters designed are evaluated by applying them to schemes for rectangular and quincunx downsampling of images. Concluding remarks are given in Section 4.9.

4.2 EXISTING METHODS FOR 2-D FREQUENCY SAMPLING

The frequency sampling method is probably the least studied among the traditional approaches to 2-D FIR filter design. *Uniform frequency sampling* methods were the first to appear. They involve sampling the 2-D frequency response of the ideal filter at the vertices of a Cartesian grid on the frequency plane, and then solving for the corresponding impulse response using a 2-D inverse DFT [Lim, 1990; Hu and Rabiner, 1972]. This method is computationally efficient. For designing a filter of size $N \times N$, it requires $O(N^3)$ operations if the inverse DFT is used, or $O(N \log_2 N)$ operations if an FFT is applied (assuming that N is a power of two in the latter case). However, if we use this method to design filters that are piecewise constant, the sharp transition between bands causes large ripples in the resulting frequency response. This is improved considerably if some transition samples are introduced. Computing optimal values for these transition samples requires linear programming methods that minimize the peak ripple [Hu and Rabiner, 1972]. This is computationally intensive due to the large number of constraint equations obtained by sampling the 2-D frequency response over a dense grid. Therefore, this is typically avoided by simply using a reasonable choice for the transition sample values, such as linear interpolation of samples. The inherent disadvantage of uniform frequency sampling is that there is *no flexibility* in the placement of frequency samples. Consequently, the filters deviate from the desired passband shape, especially for small filter sizes.

There have been efforts to improve this method by using a *nonuniform* placement of the frequency samples [Zakhor and Alvstad, 1992; Angelides, 1994; Rozwood et al., 1991; Diamessis et al., 1987]. Interestingly, the generalized frequency sampling method forms the basis for some iterative algorithms used for the design of optimal 2-D FIR filters [Kamp and Thiran, 1975; Harris and Mersereau, 1977]. The approach proposed by Diamessis, Therrien, and Rozwood, is based on an extension of Newton interpolation to 2-D [Diamessis et al., 1987]. In order to generate a Newton form in 2-D, some topological constraints must be placed on the frequency support. This particular method uses a triangular support for the frequency samples. The paper [Diamessis et al., 1987] contains only one example illustrating the design of a 2-D lowpass square-shaped filter. It is difficult to design filters with more complex shapes using such a restricted support for the frequency-sample locations.

Angelides has proposed an approach using samples on a nonuniform rectangular grid and a class of 2-D Newton-type polynomials for designing 2-D FIR filters [Angelides, 1994]. This reduces the 2-D design problem to two 1-D linear systems, as described in Chapter 2 (Section 2.5.2, Case 1). The computational complexity is $O(N^3)$.

In the method proposed by Rozwood, Therrien, and Lim [Rozwood et al., 1991], the locations of the frequency samples are constrained to be on parallel horizontal or vertical lines in the 2-D frequency plane. As described in Chapter 2 (Section 2.5.2, Case 2), this form of sampling reduces the general linear system to smaller sets of equations, thus reducing complexity. The authors give two design examples for circularly symmetric lowpass and bandpass filters, in which some samples were placed along the edges of the passband and stopband to provide greater control over the shape. These filters have band-edges which are more circular as compared with filters designed by uniform frequency sampling, but at the expense of larger peak ripples. However, there is no efficient algorithm which will automatically locate the samples satisfying the constraints of this method, and design a filter of acceptable quality. This method has a computational complexity of $O(N^4)$.

Zakhor and Alvstad have developed a broader class of nonuniform sampling strategies, in which they constrain the sample locations to be on irreducible curves in the 2-D frequency plane [Zakhor and Alvstad, 1992]. Their examples of sampling distributions include polar samples, and samples on straight lines. They have derived a theorem that provides conditions for unique interpolation. However, in general, the total number of samples required for unique interpolation exceeds the number of filter coefficients. This results in a linear least square (LLS) solution which requires $O(N^6)$ operations. The examples given include the design of circular lowpass and directional filters [Zakhor and Alvstad, 1992]. A common feature of these designs is that a few samples are placed at the intersection of the sampling lines and the edges of the passband and stopband, to shape the contours in the desired fashion.

4.3 PROPOSED 2-D NONUNIFORM FREQUENCY SAMPLING DESIGN

In the proposed 2-D filter design method, the desired frequency response is sampled at N_i points located nonuniformly in the 2-D frequency plane, where N_i is the number of independent filter coefficients [Bagchi and Mitra, 1995b; Bagchi and Mitra, 1996b]. All symmetries present in the filter impulse response are utilized so that N_i is typically much lower than the total number of filter coefficients, N^2. This reduces the design time, besides guaranteeing symmetry. The set of N_i linear equations, given by the 2-D NDFT formulation in Eq. (2.115) of Chapter 2, is then solved to obtain the filter coefficients.

As in the case of 1-D filter design, the choice of the sample values and locations depends on the particular type of filter being designed. In general, the problem of locating the 2-D frequency samples is much more complex than in the 1-D case. Our experience in designing 2-D filters with various shapes indicates that best results are obtained when the samples are placed on *contour lines* that match the desired passband shape. For example, to design a square-shaped filter, we place the samples along a set of square contour lines in the 2-D frequency plane. Note that these results agree with the filter design results reported earlier [Zakhor and Alvstad, 1992; Rozwood et al., 1991], where better control over shape was obtained by placing samples at the edges of the passband and the stopband. The total number of contours and number of samples on each contour have to be chosen carefully so as to avoid singularities. A necessary condition for nonsingularity is known [Zakhor and Alvstad, 1992, Theorem 2, p. 171], and helps to serve as a rough check. However, this condition is not sufficient to guarantee nonsingularity. This theorem asserts that if the sum of the degrees of the irreducible curves, on which the samples are placed, is small compared to the degree of the filter polynomial, then the interpolation problem becomes singular. Going back to our example of designing a square filter, it is clear that the number of square contours must be chosen appropriately with respect to the filter size.

As we locate the frequency samples along contour lines of the desired shape, the *parameters* to be chosen are: (a) the number of contours and the spacing between them, (b) the number of samples on each contour and their relative spacing, and (c) the sample values. In the following sections, we show how these parameters are chosen for filters with various shapes, such as square, circular, diamond, fan, and elliptic shapes. A common approach used is that a particular *cross-section* of the desired 2-D frequency response is approximated by 1-D analytic functions based on Chebyshev polynomials, similar to those used for 1-D FIR filter design in Chapter 3. The samples are then placed on contours that pass through the extrema of this cross-section. This will become clear when we look at design examples in the following sections.

4.4 SQUARE FILTER DESIGN

Consider the design of a square-shaped lowpass filter $h[n_1, n_2]$, with an amplitude-response specification shown in Fig. 4.1. Let the filter be of size $N \times N$, with passband edge ω_p and stopband edge ω_s, as defined in Fig. 4.1. We consider the frequency response $H(e^{j\omega_1}, e^{j\omega_2})$ to be of zero-phase, with an amplitude response $A(\omega_1, \omega_2)$ satisfying the condition

$$A(\omega_1, \omega_2) = A^*(\omega_1, \omega_2). \tag{4.1}$$

For a real impulse response $h[n_1, n_2]$, Eq. (4.1) implies that $h[n_1, n_2]$ is symmetric with respect to the origin in the spatial plane:

$$h[n_1, n_2] = h[-n_1, -n_2]. \tag{4.2}$$

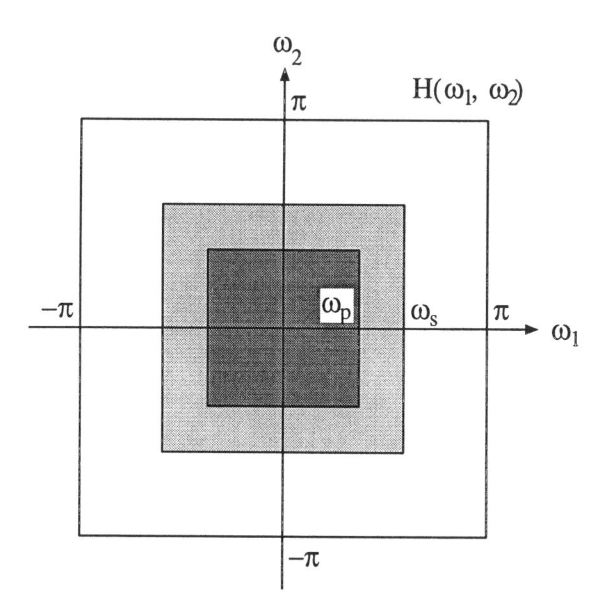

Figure 4.1. Amplitude-response specification of a square-shaped lowpass filter. Darkly shaded region: passband, lightly shaded region: transition band, unshaded region: stopband.

From Fig. 4.1, it is clear that the amplitude response exhibits a fourfold symmetry [Lim, 1990],given by

$$A(\omega_1, \omega_2) = A(-\omega_1, \omega_2) = A(\omega_1, -\omega_2). \tag{4.3}$$

In the spatial domain, this is equivalent to

$$h[n_1, n_2] = h[-n_1, n_2] = h[n_1, -n_2]. \tag{4.4}$$

Using Eqs. (4.1) and (4.3), we can express the amplitude response in the form

$$\begin{aligned} A(\omega_1, \omega_2) &= h[0,0] + \sum_{n_1=1}^{(N-1)/2} 2h[n_1, 0] \cos \omega_1 n_1 + \sum_{n_2=1}^{(N-1)/2} 2h[0, n_2] \cos \omega_2 n_2 \\ &+ \sum_{n_1=1}^{(N-1)/2} \sum_{n_2=1}^{(N-1)/2} 4h[n_1, n_2] \cos \omega_1 n_1 \cos \omega_2 n_2. \end{aligned} \tag{4.5}$$

Thus, the number of independent filter coefficients is given by

$$N_i = \frac{(N+1)^2}{4}. \tag{4.6}$$

To solve for these coefficients, we require N_i samples of $A(\omega_1, \omega_2)$ located in the first quadrant of the (ω_1, ω_2) plane.

Our design method is based on the following idea. If we take a cross-section of the 2-D amplitude response along the ω_1 axis (or ω_2 axis), then the plot looks like a 1-D lowpass filter response. An example of this cross-section is shown in Fig. 4.2. Given the 2-D filter specifications such as the support size and band-edges, we first represent the passband and stopband of this cross-section by separate analytic functions, $H_p(\omega)$ and $H_s(\omega)$, as was done for 1-D lowpass filter design in Chapter 3 (Section 3.3.1). Then we place samples along the ω_1 axis, at the extrema of these functions. In the 2-D frequency plane, the samples are placed on squares passing through these extrema. All the samples on a particular square have the same value and are evenly spaced. The total number of square contours is $(N+1)/2$. The number of samples on the kth contour starting from the origin is $(2k-1)$, for $k = 1, 2, \ldots, (N+1)/2$. This choice works well, since the total number of samples is given by the sum of the series of $(N+1)/2$ odd numbers:

$$\sum_{k=1}^{(N+1)/2} (2k-1) = \left(\frac{N+1}{2}\right)^2, \tag{4.7}$$

which, from Eq. (4.6), equals N_i.

The filter coefficients are found by solving the N_i equations obtained by sampling Eq. (4.5) at N_i points. The above design method can also be used to design square-shaped *highpass* filters. In this case, only one modification is

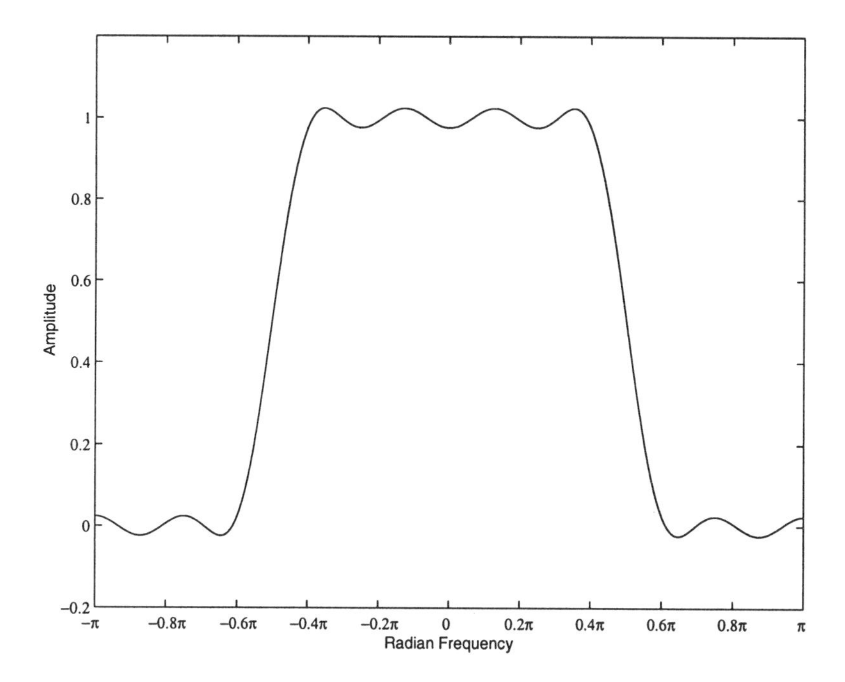

Figure 4.2. Cross-section of desired square filter amplitude response along the ω_1 axis.

required—the cross-section of the 2-D amplitude response along the ω_1 axis (or ω_2 axis) has to be approximated by a 1-D highpass response.

The following example illustrates the design method.

Example 4.1

Consider the design of a square lowpass filter with the following specifications:

Support size $= 9 \times 9$,
Passband edge $\omega_p = 0.35\pi$, Stopband edge $\omega_s = 0.65\pi$.

Since only 25 of the 81 filter coefficients are independent, we require 25 samples located in the first quadrant of the 2-D frequency plane. The total number of square contours is five. The cross-section of the 2-D amplitude response along the ω_1 axis is represented by 1-D analytic functions, as shown in Fig. 4.3. Then, the location and values of the five samples along the ω_1 axis are obtained by sampling these functions at the points marked on this figure. Finally, the remaining samples are placed on five successive squares passing through these points, as illustrated in Fig. 4.4(b). These samples lead to linear equations which are solved to determine the filter coefficients. The amplitude response

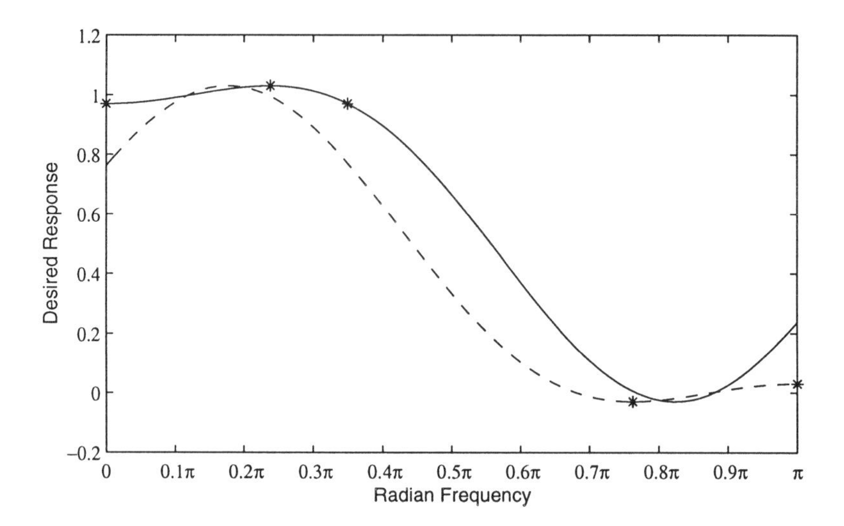

Figure 4.3. Generation of analytic functions for square filter designed by the NDFT method. The cross-section of the 2-D amplitude response along the ω_1 axis is approximated by $H_p(\omega)$ (solid line) in the passband, and $H_s(\omega)$ (dashed line) in the stopband. These functions are sampled at the locations denoted by "$*$".

and contour plot of the resulting square filter are shown in Figs. 4.4(a) and (b). Clearly, the passband has the desired square shape.

4.4.1 Comparison With Separable Square Filters

The frequency response of a square filter, shown in Fig. 4.1, is essentially a separable one. We can design a separable 2-D square filter $h[n_1, n_2]$ by first designing two 1-D lowpass filters $h_1[n]$ and $h_2[n]$, and then simply taking their product along orthogonal directions, as given by

$$h[n_1, n_2] = h_1[n_1]h_2[n_2]. \tag{4.8}$$

In the frequency domain, this is equivalent to

$$A(\omega_1, \omega_2) = A_1(\omega_1)A_2(\omega_2), \tag{4.9}$$

where $A_1(\omega_1)$ and $A_2(\omega_2)$ are the amplitude functions of the 1-D filters $h_1[n]$ and $h_2[n]$, respectively.

Separable filters are widely used in practice because of their design simplicity and ease of implementation. However, they suffer from the following problem. If δ_{p_1}, δ_{s_1} and δ_{p_2}, δ_{s_2} are the peak ripples in the passband and stopband of the two 1-D filters, respectively, then the resulting 2-D filter can have peak ripples as large as $(\delta_{p_1} + \delta_{p_2})$ in the passband, and $\max(\delta_{s_1}, \delta_{s_2})$ in the stopband.

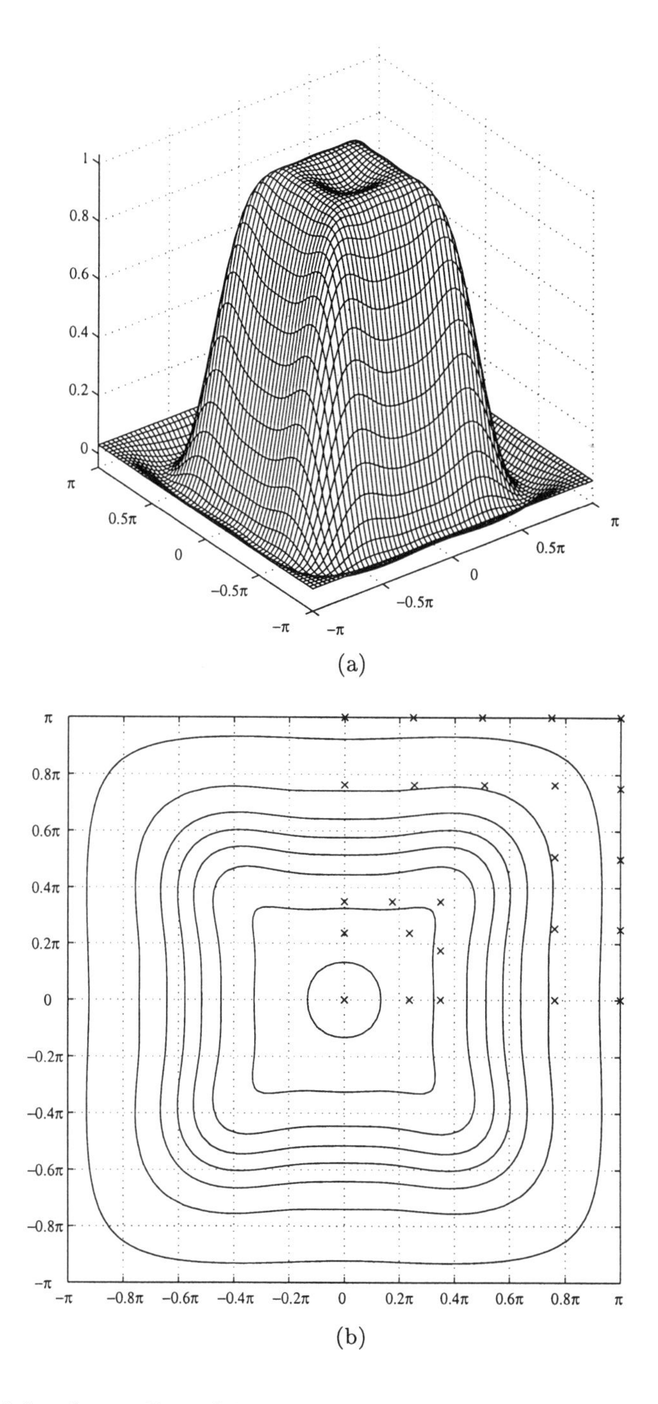

Figure 4.4. Square filter of size 9×9 designed by the NDFT method. (a) Plot of amplitude response $A(\omega_1, \omega_2)$ with normalized frequencies $(\omega_1/2\pi)$ and $(\omega_2/2\pi)$. (b) Contour plot. The sample locations are denoted by "×".

Thus, the 2-D filter might have large passband ripple, which is undesirable. In such cases, much better results can be obtained by utilizing the N^2 degrees of freedom available in designing a nonseparable filter, instead of the $2N$ degrees used in a separable design.

We now compare the nonseparable square filter in Example 4.1 to a separable filter designed with the same specifications. The Parks-McClellan algorithm is used to design the corresponding 1-D lowpass filter with length $N = 9$, and band-edges as specified in Example 4.1. Figs. 4.5(a) and (b) show the amplitude response and contour plot of the separable square filter. Note the large ripple obtained in the passband. The performance comparison is shown in Table 4.1. We denote the peak ripples obtained in the passband and stopband by δ_p and δ_s, respectively. Note that the peak passband ripple obtained by our method is nearly a third of that present in the separable filter.

Table 4.1. Performance comparison for square filter design.

Method	δ_p	δ_s
NDFT	0.0322	0.0471
Separable Design	0.1116	0.0579

4.5 CIRCULARLY SYMMETRIC FILTER DESIGN

Consider next, the design of a circularly symmetric lowpass filter of size $N \times N$, which has an amplitude-response specification with passband edge ω_p and stopband edge ω_s, as defined in Fig. 4.6. Such filters exhibit a fourfold symmetry, as given in Eq. (4.4). Therefore the frequency response $A(\omega_1, \omega_2)$ can be expressed as shown in Eq. (4.5). The number of independent filter coefficients N_i is given by Eq. (4.6).

As for square filter design, we require samples of the desired response at N_i points in the first quadrant of the frequency plane. These samples are placed on circular contours. The cross-section of the 2-D amplitude response along the ω_1 axis (or ω_2 axis) looks like a 1-D lowpass response, and is represented by analytic functions as before. The $(N+1)/2$ samples along the ω_1 axis are placed at the extrema of these functions. We can view the process of generating the remaining samples as a rotation of the samples located along the ω_1 axis. For ease of understanding, this is illustrated in Example 4.2.

Circularly symmetric highpass filters can also be designed using this method, by generating the 1-D analytic functions so as to get a highpass response along the ω_1 axis.

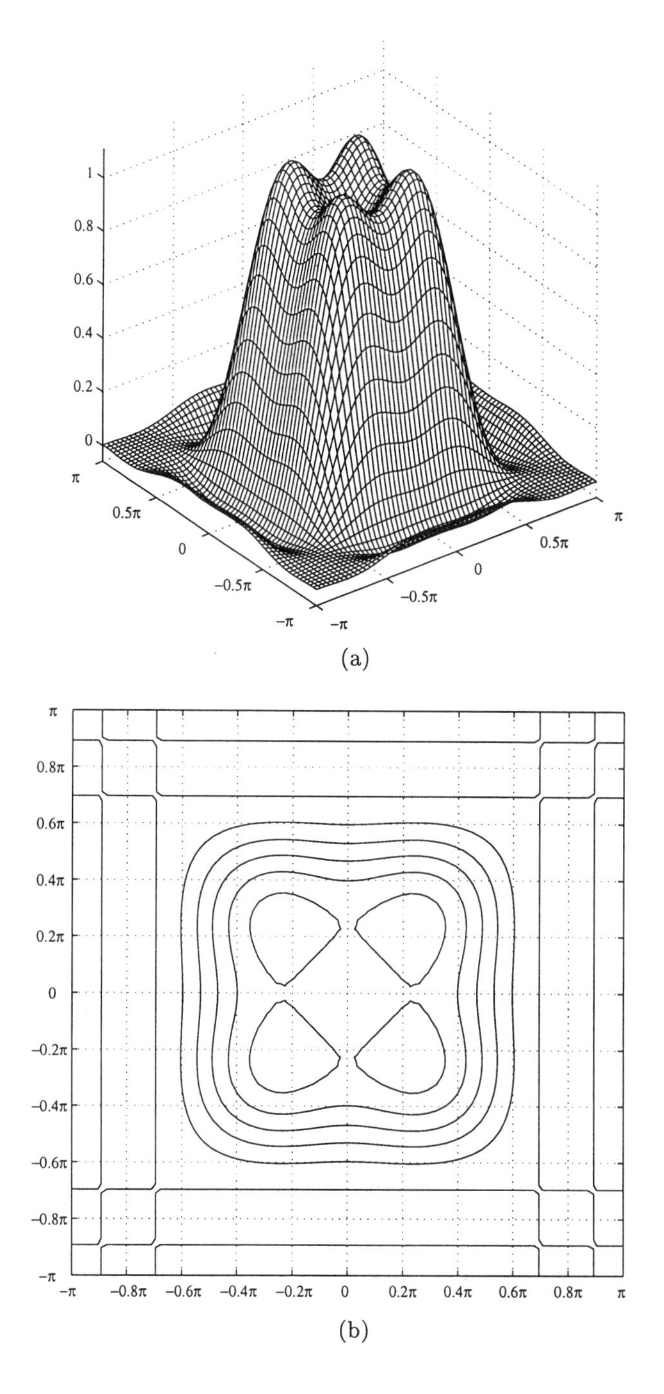

Figure 4.5. Separable square filter of size 9 × 9. (a) Amplitude response. (b) Contour plot.

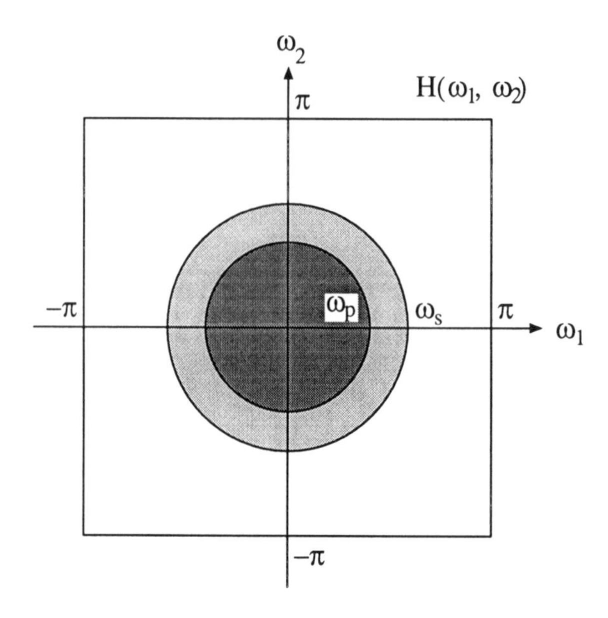

Figure 4.6. Amplitude-response specification of a circularly symmetric lowpass filter. Darkly shaded region: passband, lightly shaded region: transition band, unshaded region: stopband.

Example 4.2

Consider the design of a circularly symmetric lowpass filter with the following specifications:

Support size $= 15 \times 15$,
Passband edge $\omega_p = 0.4\pi$, Stopband edge $\omega_s = 0.6\pi$.

The 64 samples required are placed on circular contours in the first quadrant of the (ω_1, ω_2) plane, as shown in Fig. 4.7(b). The values and locations of the eight samples along the ω_1 axis are first obtained by generating the 1-D response along this axis. These samples are then used to generate $1, 3, 4, 5, 6, 7, 8, 9$ samples placed on eight respective circles starting from the origin. The rest of the samples in the upper right corner near $(\omega_1, \omega_2) = (\pi, \pi)$ are placed on circular arcs uniformly spaced from each other. As can be seen in Fig. 4.7(b), there are $6, 5, 4, 3, 2, 1$ samples on six successively smaller arcs. Such a choice of samples works well with a number of other filters designed. For a filter of size $N \times N$, we place $1, 3, 4, 5, \ldots, (N+3)/2$ samples on $(N+1)/2$ circles, and $(N-3)/2, (N-5)/2, \ldots, 3, 2, 1$ samples on $(N-3)/2$ circular arcs, going radially outward from the origin. The samples on a particular circle or arc are placed at equal angles with respect to the origin. Figs. 4.7(a) and (b) show the frequency response and contour plot of the 15×15 circular lowpass filter.

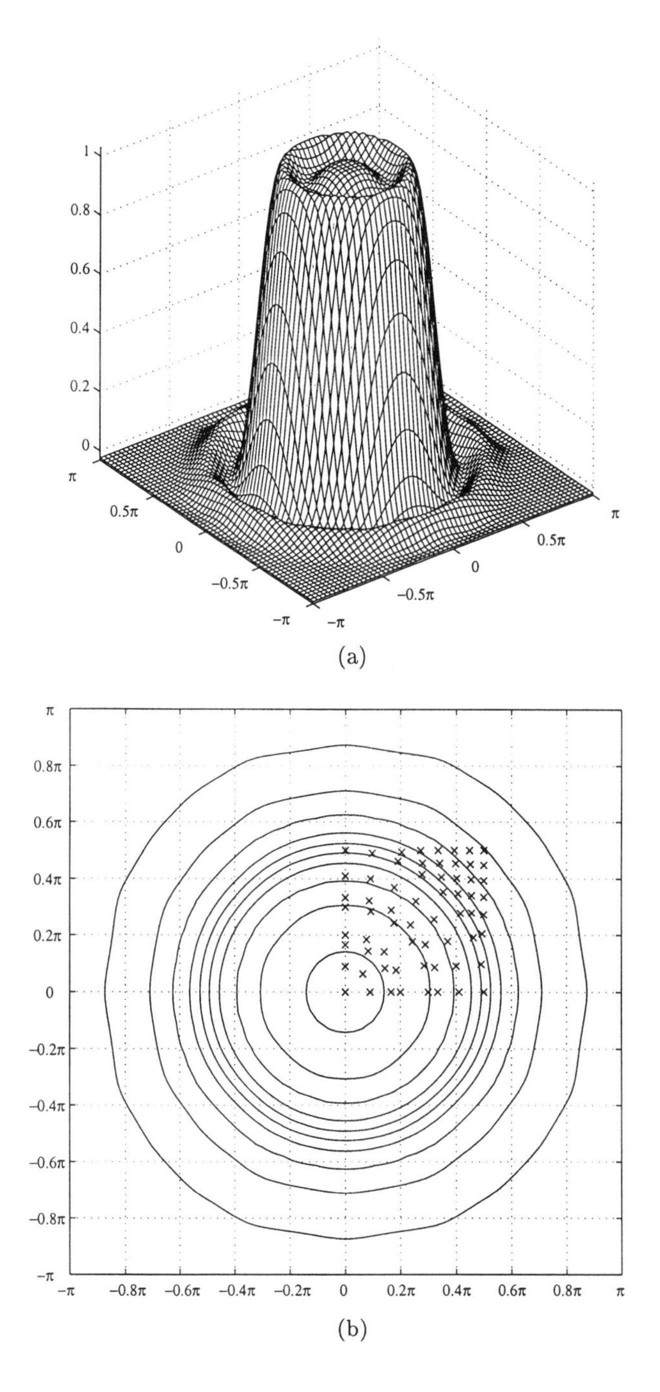

Figure 4.7. Circular lowpass filter of size 15×15 designed by the NDFT method. (a) Amplitude response. (b) Contour plot. The sample locations are denoted by "$\times$".

The filter exhibits nearly equiripple characteristics and a high degree of circular symmetry.

4.5.1 *Comparison With Other Design Methods*

For the purpose of comparison, we consider five other methods for circularly symmetric 2-D filter design:

(1) Uniform frequency sampling [Lim, 1990].

(2) Modified uniform frequency sampling [Lightstone et al., 1994].

(3) Nonuniform frequency sampling using a linear least square (LLS) approach [Zakhor and Alvstad, 1992].

(4) McClellan frequency transformation [McClellan, 1973].

(5) Frequency transformation proposed by Hazra and Reddy [Hazra and Reddy, 1986].

We use the above methods to design a circular lowpass filter with the same specifications used in Example 4.2, and compare their performances in Table 4.2. The amplitude response and contour plots of the filters designed by Methods 1, 2, 4 and 5 are shown in Figs. 4.8–4.11, respectively. The corresponding plots for Method 3 have been presented earlier [Zakhor and Alvstad, 1992, Fig. 6, p. 175]. The NDFT method gives a good combination of low peak ripples and circular contour shapes.

Method 1 is the traditional uniform frequency sampling technique that uses sample values of unity in the passband and zeros in the stopband. The values of the samples in the transition band are chosen by linear interpolation from zero to unity. It is difficult to control the shape of the passband and stopband contours in this method.

In Method 2, the frequency samples are uniformly spaced as in Method 1. However, the sample values are obtained by uniformly sampling a 2-D analytic

Table 4.2. Performance comparison for circular filter design.

Method	δ_p	δ_s
NDFT	0.0324	0.0315
Uniform Sampling	0.0393	0.0519
Modified Uniform Sampling	0.0245	0.0244
Nonuniform LLS	0.0360	0.0430
McClellan Transformation	0.0238	0.0238
Hazra-Reddy Transformation	0.0587	0.0587

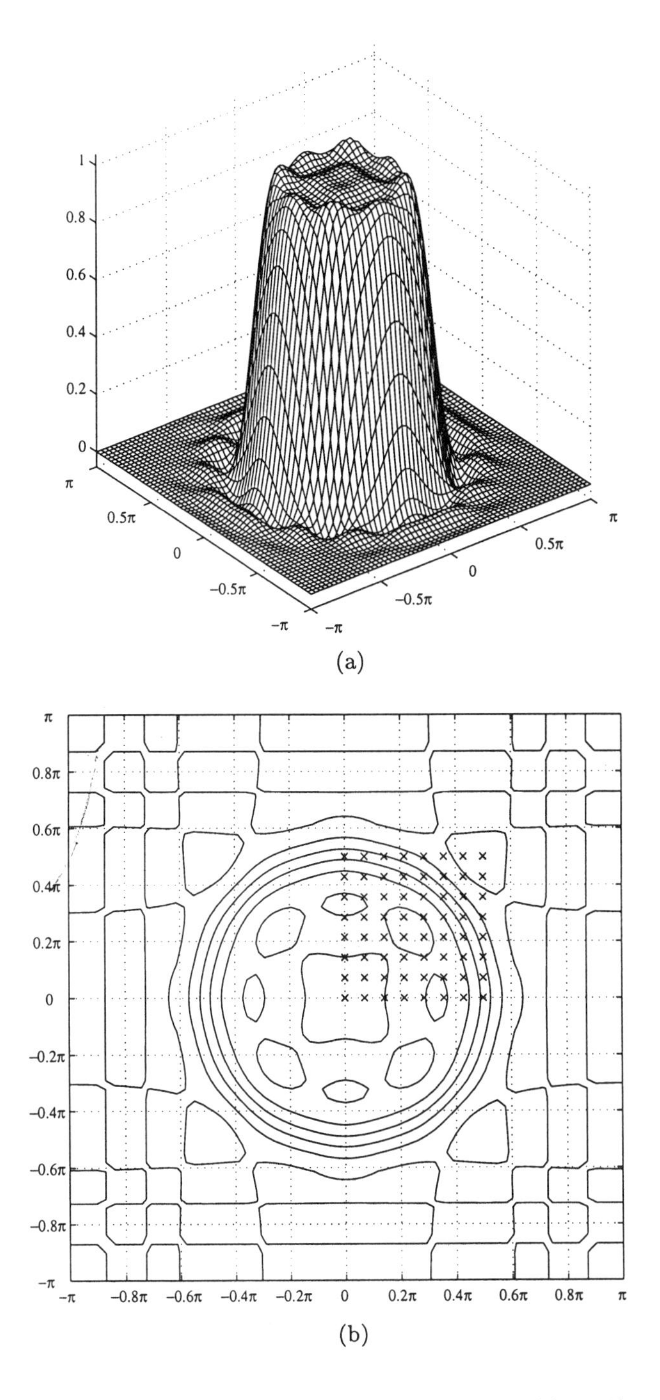

Figure 4.8. Circular lowpass filter of size 15×15 designed by uniform frequency sampling. (a) Amplitude response. (b) Contour plot. The sample locations are denoted by "$\times$".

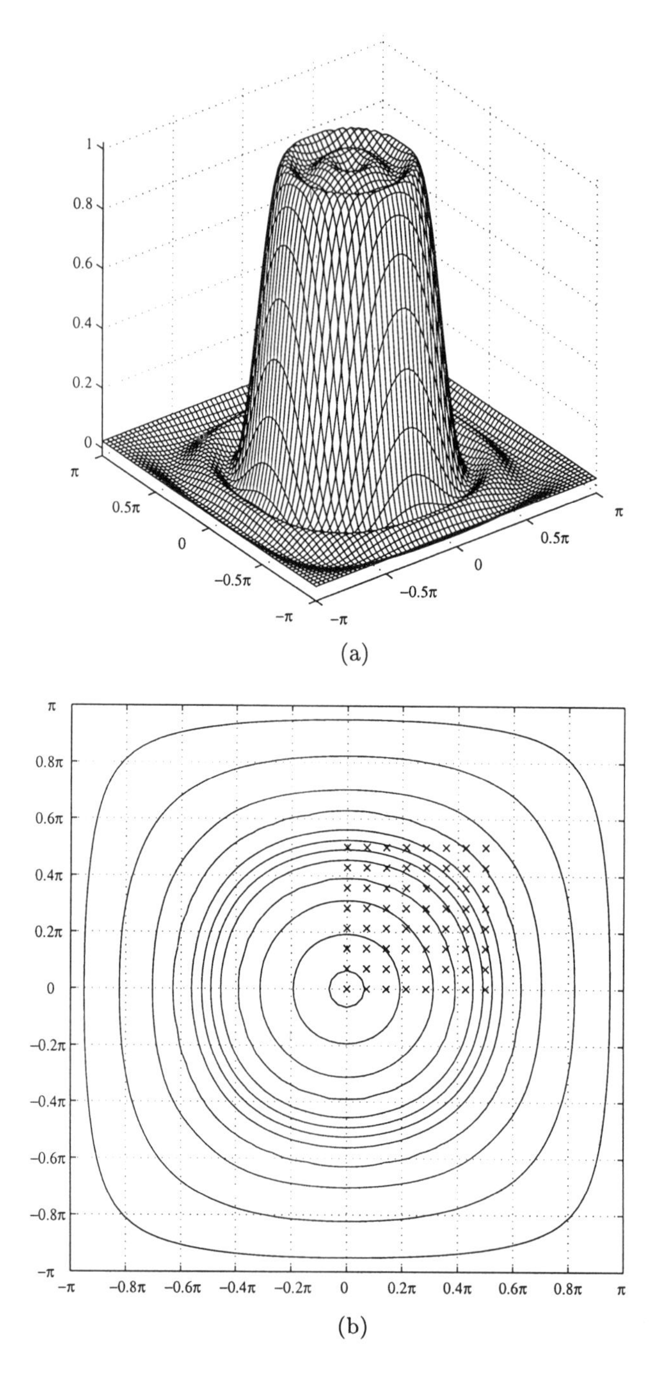

Figure 4.9. Circular lowpass filter of size 15×15 designed by modified uniform frequency sampling. (a) Amplitude response. (b) Contour plot. The sample locations are denoted by "$\times$".

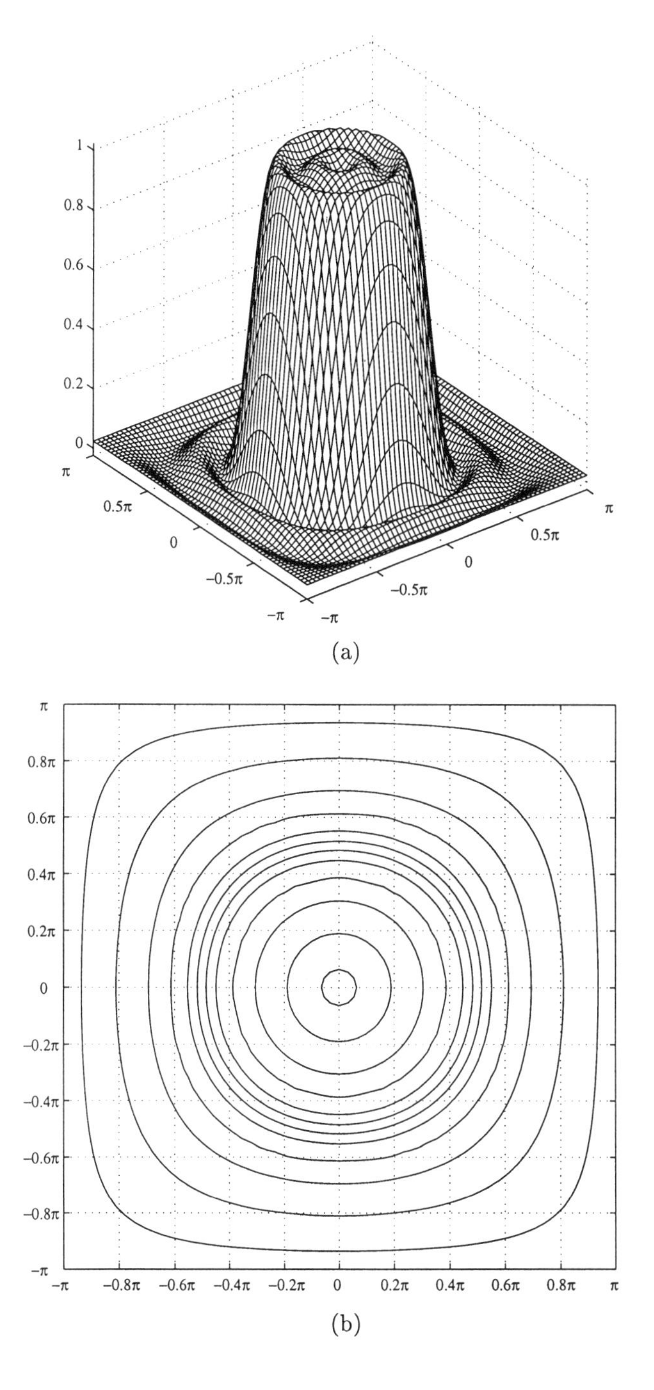

Figure 4.10. Circular lowpass filter of size 15 × 15 designed by the McClellan frequency transformation method. (a) Amplitude response. (b) Contour plot.

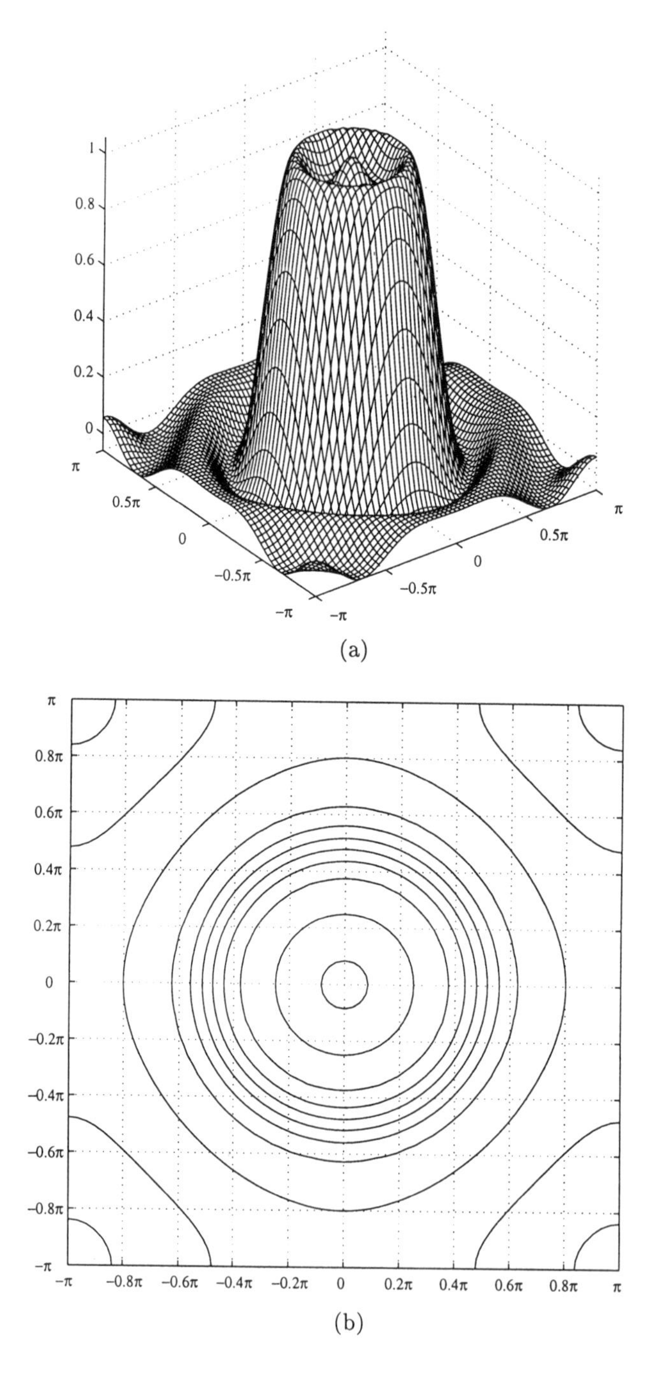

Figure 4.11. Circular lowpass filter of size 15×15 designed by the Hazra-Reddy frequency transformation method. (a) Amplitude response. (b) Contour plot.

function generated by applying the McClellan frequency transformation to the 1-D analytic functions used for 1-D filter design in Chapter 3 (Section 3.3.1). Methods 2 and 4 produce similar filters, whose contours deviate from the desired circular shape as the frequency increases from 0 to π. This method produces filters which are nearly the same as those obtained by the McClellan frequency transformation method. Both these methods produce increasingly square contours for higher frequencies.

In Method 3, nonuniform frequency sampling is used with a least square approach, as discussed earlier in Section 4.2. For comparison, we refer to an example [Zakhor and Alvstad, 1992, p. 173] that uses the same specifications as in Example 4.2. Here, a large number of samples were placed on 15 lines having slopes ± 1 in the frequency plane. The authors note that this method results in contours which are more circular than for uniform frequency sampling in Method 1. The peak ripples reported are shown in Table 4.2.

In Method 4, a 2-D filter is designed by applying the McClellan frequency transformation to a 1-D filter. This transformation is of size 3×3, and is given by

$$\cos\omega = T(\omega_1, \omega_2) = -\frac{1}{2} + \frac{1}{2}\cos\omega_1 + \frac{1}{2}\cos\omega_2 + \frac{1}{2}\cos\omega_1\cos\omega_2. \tag{4.10}$$

For our example, we designed a 1-D filter with length $N = 15$ and band-edges as given in Example 4.2, using the Parks-McClellan algorithm. As with Method 2, this produces increasingly square contours for higher frequencies.

Method 5 is an improved frequency transformation technique introduced by Hazra and Reddy. Given the band-edges of the 2-D filter, a transformation of size 3×3 is designed [Hazra and Reddy, 1986]. This produces 2-D filters which are much more circular than those produced by the McClellan transformation, but at the expense of higher peak ripples for the same filter specifications.

4.6 DIAMOND FILTER DESIGN

Diamond filters find important practical applications as prefilters for quincunxially sampled data, and in interlaced-to-noninterlaced scanning converters for television signals [Tonge, 1981]. A diamond filter has an amplitude-response specification with passband edge ω_p and stopband edge ω_s, as defined in Fig. 4.12. In other words, the diagonal line $\omega_1 = \omega_2$ in the frequency plane intersects the passband edge at (ω_p, ω_p) and the stopband edge at (ω_s, ω_s). A diamond filter exhibits an eightfold symmetry in the frequency domain [Lim, 1990], given by

$$A(\omega_1, \omega_2) = A(-\omega_1, \omega_2) = A(\omega_1, -\omega_2) = A(\omega_2, \omega_1). \tag{4.11}$$

Equivalently, the impulse response $h[n_1, n_2]$ satisfies the relation

$$h[n_1, n_2] = h[-n_1, n_2] = h[n_1, -n_2] = h[n_2, n_1]. \tag{4.12}$$

Besides, a diamond filter is a 2-D half-band filter. Its amplitude response $A(\omega_1, \omega_2)$ is symmetric about the point $(\omega_1, \omega_2, A(\omega_1, \omega_2)) = (\pi/2, \pi/2, 0.5)$ in

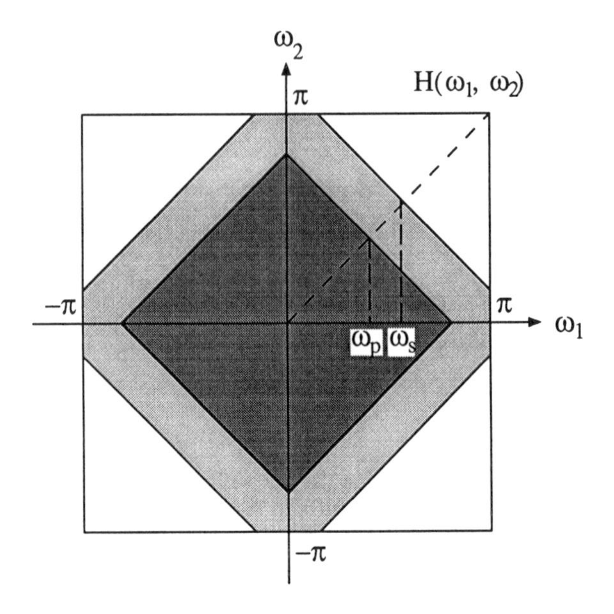

Figure 4.12. Amplitude-response specification of a diamond filter. Darkly shaded region: passband, lightly shaded region: transition band, unshaded region: stopband.

frequency space:

$$A(\omega_1, \omega_2) + A(\pi - \omega_1, \pi - \omega_2) = 1. \tag{4.13}$$

This implies that the impulse response has alternating zeros so that the non-zero coefficients form a quincunx-like lattice. Thus, we get

$$h[n_1, n_2] = \begin{cases} 0, & n_1 + n_2 = \text{even}, \\ 0.5, & n_1 = n_2 = 0. \end{cases} \tag{4.14}$$

For example, the arrangement of points in the impulse response of a 9×9 half-band filter is illustrated in Fig. 4.13. On account of the properties in Eq. (4.12) and (4.14), the number of independent coefficients [Yoshida et al., 1990] in a filter of size $N \times N$ is reduced to

$$N_i = \left\lfloor \frac{P+1}{2} \right\rfloor \left\lfloor \frac{P+2}{2} \right\rfloor, \tag{4.15}$$

where

$$P = \frac{N-1}{2}. \tag{4.16}$$

The N_i independent points in the impulse response lie in a wedge-shaped region below the diagonal line $n_1 = n_2$ in the first quadrant of the (n_1, n_2)

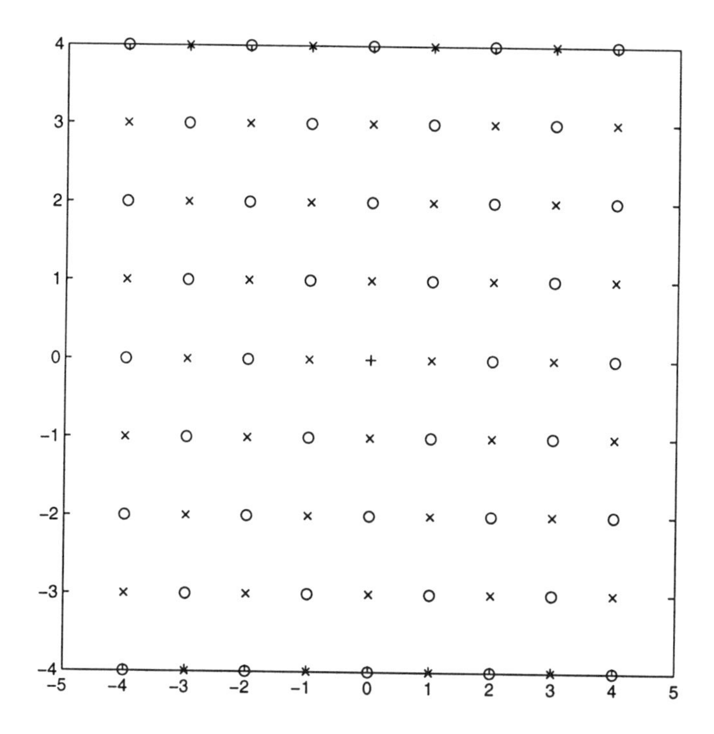

Figure 4.13. Arrangement of points in the impulse response of a 9×9 half-band FIR filter. Zero-valued samples are denoted by "o", and nonzero samples are denoted by "×". The value of the sample at "+" is 0.5.

spatial plane. By using Eqs. (4.12) and (4.14), the amplitude response of a diamond filter [Yoshida et al., 1990] is expressed as

$$\begin{aligned} A(\omega_1, \omega_2) &= 0.5 + \sum_{n_1=1}^{\lfloor (P+1)/2 \rfloor} 2h[2n_1 - 1, 0]\{\cos(2n_1 - 1)\omega_1 + \cos(2n_1 - 1)\omega_2\} \\ &\quad + \sum_{n_1=1}^{\lfloor (P+1)/2 \rfloor} \sum_{n_2=1}^{\lfloor P/2 \rfloor} 4h[2n_1 - 1, 2n_2]\{\cos(2n_1 - 1)\omega_1 \cos(2n_2)\omega_2 \\ &\quad + \cos(2n_2)\omega_1 \cos(2n_1 - 1)\omega_2\}. \end{aligned} \tag{4.17}$$

Due to the eightfold symmetry and the half-band nature of the filter, the only independent part of the amplitude response is a triangular area within the passband, as shown in Fig. 4.14. In our design method, N_i samples are placed within this region of the frequency plane. If we take a cross-section of $A(\omega_1, \omega_2)$ along the diagonal line $\omega_1 = \omega_2$, the plot looks like a 1-D half-band lowpass response. We approximate the passband of this response by a 1-D function $H_p(\omega)$, as used for 1-D half-band lowpass filter design in Chapter 3

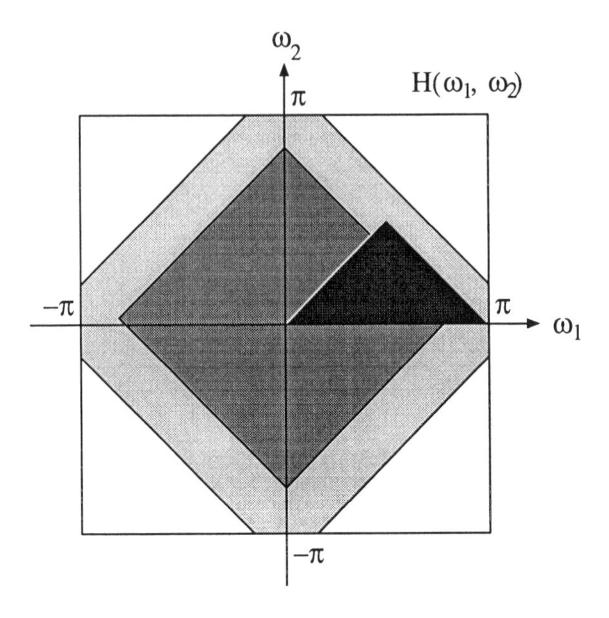

Figure 4.14. Amplitude response of a diamond filter. The darkly shaded region is the only independent part, because of eightfold symmetry and the half-band nature of the filter.

(Section 3.3.2). The order P of the corresponding Chebyshev polynomial $T_p(x)$ is $(N-1)/2$. The samples are then placed on $(N-1)/2$ lines of slope -1, that pass through the extrema of $H_p(\omega)$. All samples on a particular line have the same value and are evenly spaced. The number of samples on successive lines, as we go away from the origin, is given in Table 4.3, for filter sizes from 7×7 to 31×31. This range of filter sizes is large enough to cover the needs of most practical applications. The given distribution of samples has been found to work well for various choices of the band-edges. Note that if there is only one sample to be placed on a particular contour, it is placed on the ω_1 axis. Finally, the N_i samples are used to solve for the impulse response coefficients in Eq. (4.17).

The proposed design method produces diamond filters of high quality, with low peak ripple and better passband shape as compared with filters produced by other existing design methods. A comparison between these methods is presented in Section 4.6.1.

Example 4.3

Consider the design of a diamond filter with the following specifications:

Support size $= 9 \times 9$,
Passband edge $\omega_p = 0.36\pi$, Stopband edge $\omega_s = 0.64\pi$.

Table 4.3. Distribution of samples for diamond filter design. The last column shows the number of samples placed on $(N-1)/2$ successive contours for designing a filter of size $N \times N$, which has N_i independent coefficients.

N	N_i	Number of samples on successive contours
7	4	$1, 2, 1$
9	6	$1, 1, 2, 2$
11	9	$1, 1, 2, 3, 2$
13	12	$1, 1, 2, 3, 3, 2$
15	16	$1, 1, 2, 3, 3, 4, 2$
17	20	$1, 1, 2, 3, 3, 4, 4, 2$
19	25	$1, 1, 2, 3, 3, 4, 4, 4, 3$
21	30	$1, 1, 2, 3, 3, 3, 4, 4, 5, 4$
23	36	$1, 1, 2, 3, 3, 3, 4, 4, 5, 6, 4$
25	42	$1, 1, 2, 3, 3, 4, 4, 4, 5, 6, 6, 5$
27	49	$1, 1, 2, 3, 3, 4, 4, 4, 5, 6, 6, 6, 5$
29	56	$1, 1, 2, 3, 3, 4, 4, 4, 5, 6, 6, 6, 7, 5$
31	64	$1, 1, 2, 3, 3, 4, 4, 4, 5, 6, 6, 6, 7, 7, 6$

Only 6 of the 81 filter coefficients are independent. Thus, 6 samples are placed as shown in Fig. 4.15(b), on lines that follow the diamond shape. The diagonal cross-section of the 2-D amplitude response is represented by a function $H_p(\omega)$, that has 4 extrema. Samples are placed on 4 lines passing through these extrema. The number of samples on these lines are $1, 1, 2, 2$, respectively, as given in Table 4.3. Figs. 4.15(a) and (b) show the amplitude response and contour plot of the resulting diamond filter.

As another example, the choice of sample locations used for designing a 13×13 diamond filter is shown superimposed on its contour plot in Fig. 4.16.

4.6.1 Comparison With Other Design Methods

We now consider three existing methods for designing diamond filters:

(1) Frequency transformation [Lim, 1990].

(2) Method proposed by Bamberger and Smith [Bamberger and Smith, 1992].

(3) Method proposed by Chen and Vaidyanathan [Chen and Vaidyanathan, 1993].

We use each of these methods to design a diamond filter with the same specifications as given in Example 4.3. Their performance comparison is shown in Table 4.4. The amplitude response and contour plots of the filters are shown

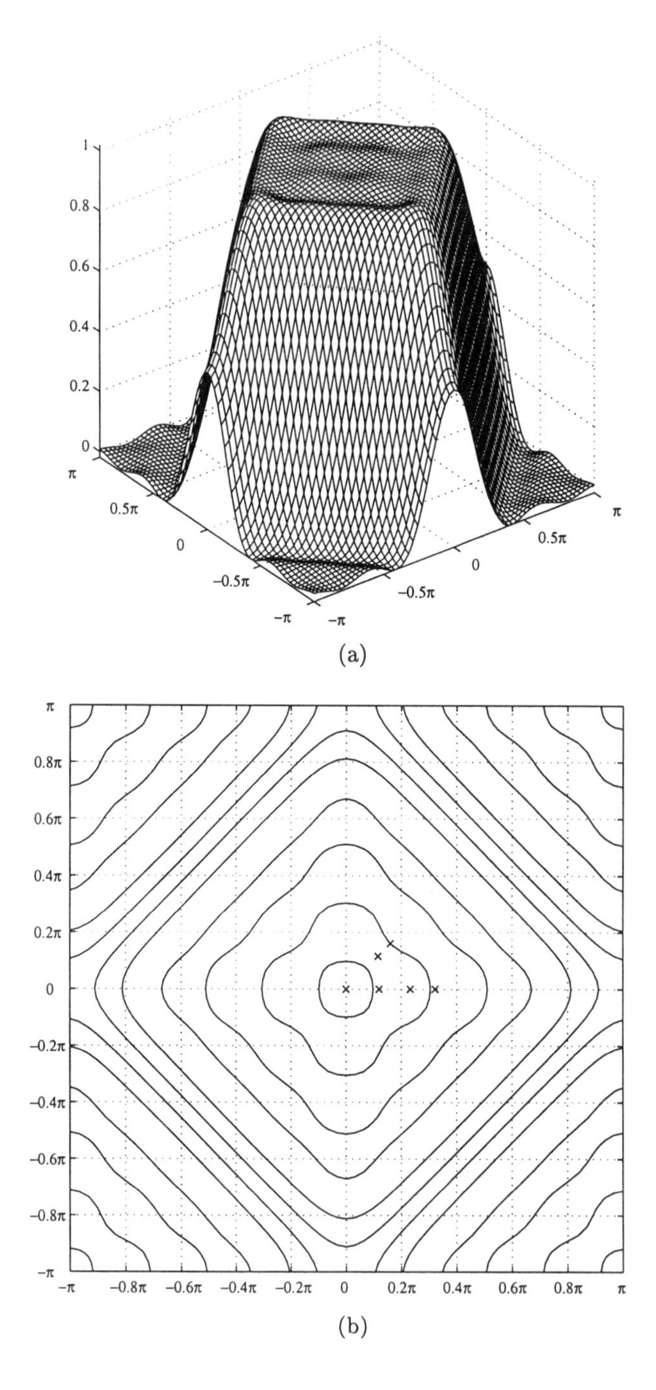

Figure 4.15. Diamond filter of size 9×9 designed by the NDFT method. (a) Amplitude response. (b) Contour plot. The sample locations are denoted by "×".

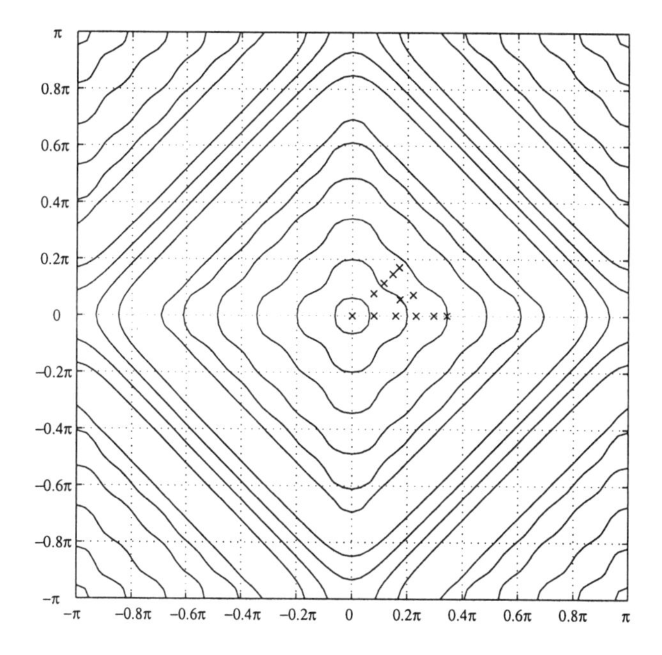

Figure 4.16. Contour plot of a diamond filter of size 13 × 13 designed by the NDFT method. The sample locations are denoted by "×".

in Figs. 4.17–4.19. From Table 4.4, we can see that the NDFT method gives the lowest peak ripples among these methods as well as good contour shapes.

In the frequency transformation method, a 1-D half-band lowpass filter is transformed to a 2-D diamond filter. The simplest transformation used to obtain the desired diamond-shaped contours is the following one of size 3 × 3:

$$\cos\omega = T(\omega_1, \omega_2) = \frac{1}{2}\cos\omega_1 + \frac{1}{2}\cos\omega_2. \tag{4.18}$$

Table 4.4. Performance comparison for diamond filter design.

Method	δ_p	δ_s
NDFT	0.0189	0.0184
Frequency Transformation	0.0636	0.0636
Bamberger-Smith	0.1085	0.1084
Chen-Vaidyanathan	0.0292	0.0281

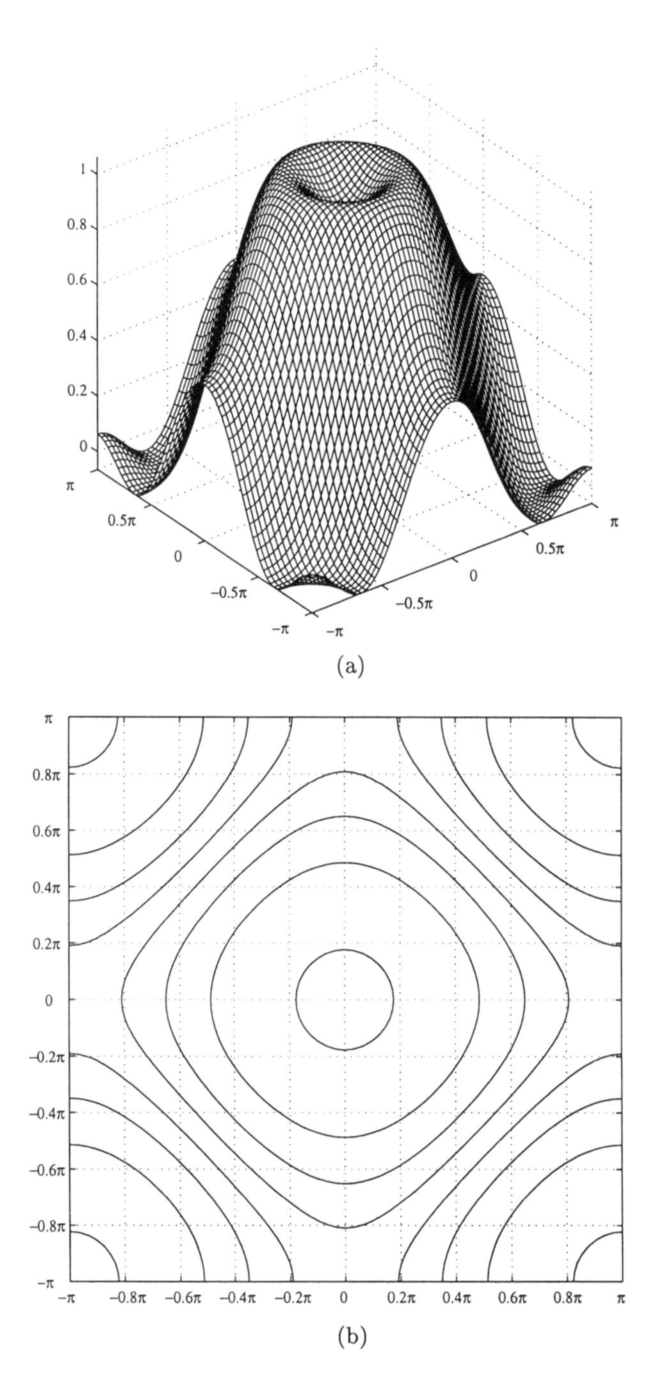

Figure 4.17. Diamond filter of size 9×9 designed by the frequency transformation method. (a) Amplitude response. (b) Contour plot.

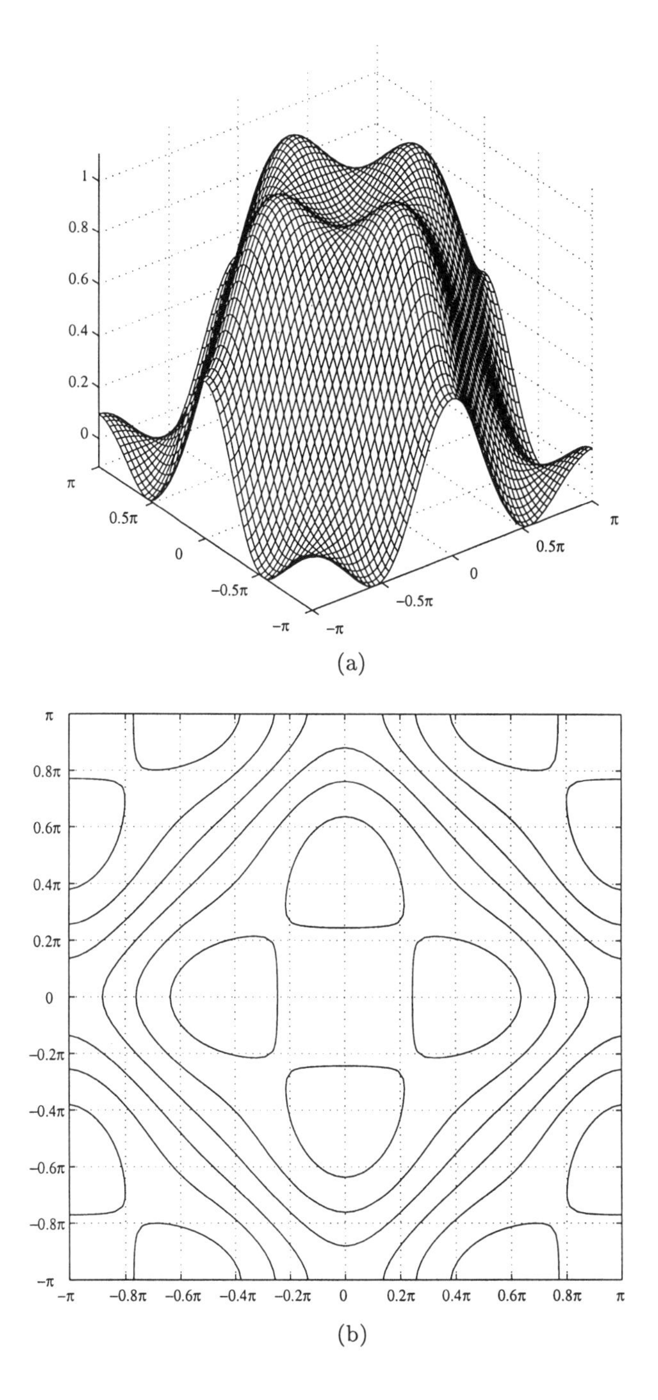

Figure 4.18. Diamond filter of size 9×9 designed by the method proposed by Bamberger and Smith. (a) Amplitude response. (b) Contour plot.

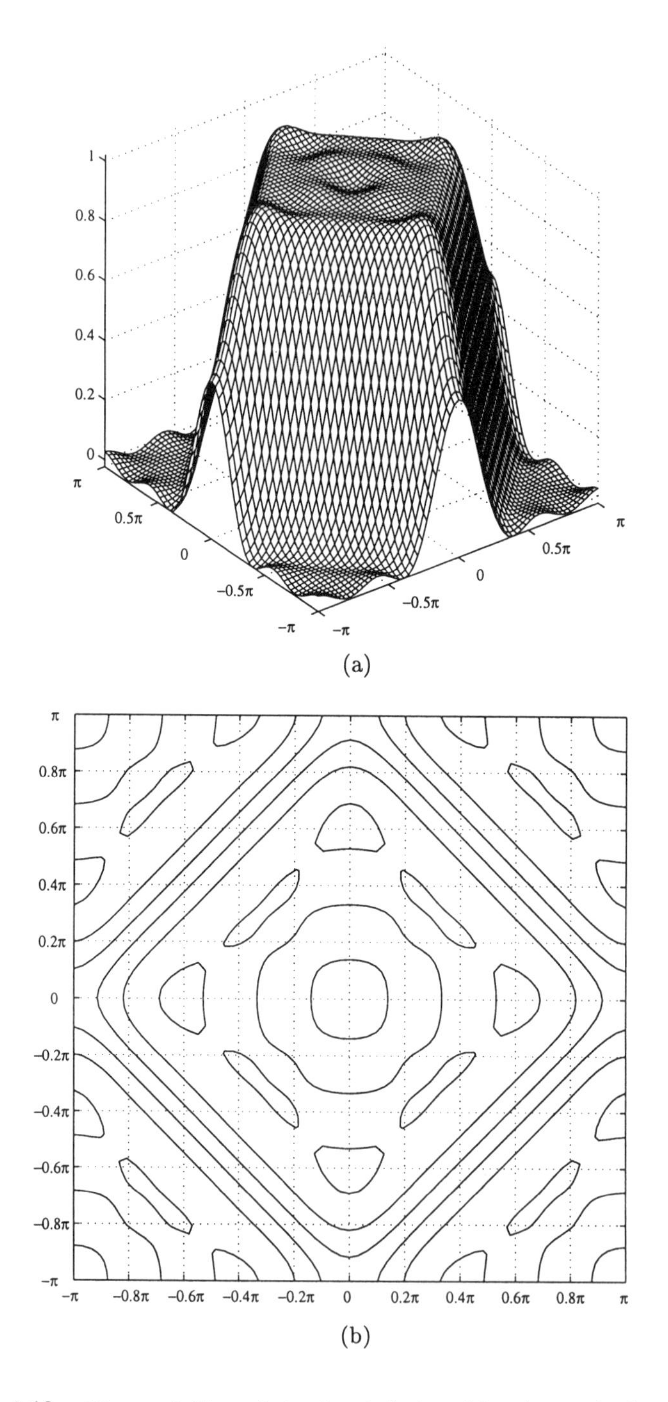

Figure 4.19. Diamond filter of size 9 × 9 designed by the method proposed by Chen and Vaidyanathan. (a) Amplitude response. (b) Contour plot.

The contours of the resulting 2-D filter are more circular rather than being diamond-shaped. The shape of the contours can be improved by using a higher order transformation, but this also increases the filter size considerably. If the transformation is of size $(2P+1) \times (2Q+1)$, and the 1-D filter is of length $(2M+1)$, then the resulting 2-D filter has size $(2PM+1) \times (2QM+1)$. Thus, in our example, if we use a 5×5 transformation and the same 1-D filter of length 9, then the 2-D filter size becomes 17×17. So this is uneconomical.

In Method 2, a diamond filter is designed by rotating a checkerboard-shaped filter through an angle of 45 degrees [Bamberger and Smith, 1992]. This checkerboard filter is the sum of two square-shaped filters,

$$C(z_1, z_2) = H_0(z_1)H_0(z_2) + H_1(z_1)H_1(z_2), \tag{4.19}$$

where $H_0(z)$ is a 1-D half-band lowpass filter, and $H_1(z) = H_0(-z)$ is the corresponding highpass filter. Although the shape of the contours is better than with frequency transformation, the ripple is too large.

Finally, we consider the method proposed by Chen and Vaidyanathan [Chen and Vaidyanathan, 1993]. This method can be used to design a general class of M-dimensional filters with arbitrary parallelepiped-shaped passbands. However, we shall only mention how it can be used to design a 2-D diamond filter of size $N \times N$. A 1-D half-band lowpass filter $h[n]$ of length $2N-1$ is designed, and used to form a separable 2-D filter,

$$h_s[n_1, n_2] = h[n_1]h[n_2]. \tag{4.20}$$

Then, this separable filter is downsampled by a quincunx matrix,

$$\hat{\mathrm{M}} = \begin{bmatrix} 1 & 1 \\ -1 & 1 \end{bmatrix}. \tag{4.21}$$

to generate the diamond filter $h[n_1, n_2]$ as follows:

$$h[n_1, n_2] = 2h_s[n_1 + n_2, n_2 - n_1]. \tag{4.22}$$

The resulting diamond filter has a good passband shape as well as low peak ripples, comparable to the results given by our method.

4.6.2 Fan Filter Design

A 90-degree fan filter is a 2-D half-band filter whose amplitude-response specification is shown in Fig. 4.20. Such filters possess directional sensitivity, which is important in applications such as the processing of geoseismic data. Comparing Figs. 4.12 and 4.20, we can see that the fan filter amplitude response is obtained by shifting the diamond amplitude response by π along the ω_2 axis. Thus, if $h_d[n_1, n_2]$ is the impulse response of a diamond filter, we can obtain the corresponding fan filter by taking the fan impulse response to be

$$h[n_1, n_2] = (-1)^{n_2} h_d[n_1, n_2]. \tag{4.23}$$

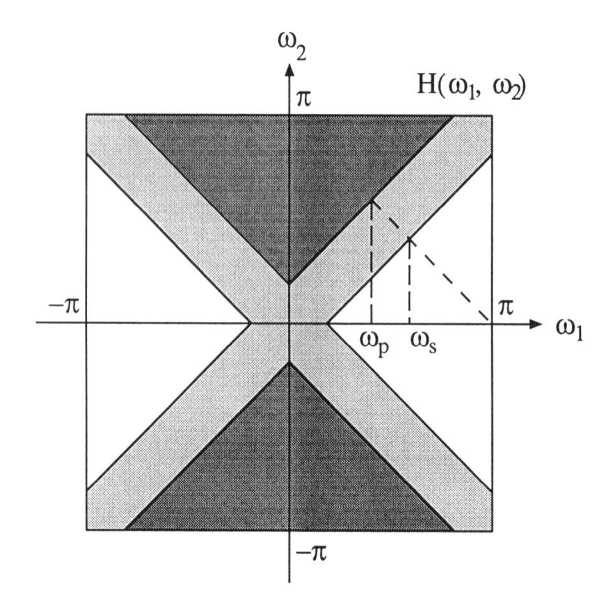

Figure 4.20. Amplitude-response specification of a fan filter. Darkly shaded region: passband, lightly shaded region: transition band, unshaded region: stopband.

Since the diamond filter satisfies the symmetry relation: $h_d[n_1, n_2] = h_d[n_2, n_1]$, we get the following antisymmetry relation for the fan filter from Eq. (4.23):

$$h[n_1, n_2] = -h[n_2, n_1], \qquad (n_1, n_2) \neq (0, 0). \tag{4.24}$$

The fan filter also exhibits a fourfold symmetry as given in Eq. (4.4). Combining this with Eq. (4.24) and the half-band condition in Eq. (4.14), we can express the amplitude response of a fan filter in the form

$$\begin{aligned} A(\omega_1, \omega_2) &= 0.5 + \sum_{n_1=1}^{\lfloor (P+1)/2 \rfloor} 2h[2n_1 - 1, 0]\{\cos(2n_1 - 1)\omega_1 - \cos(2n_1 - 1)\omega_2\} \\ &\quad + \sum_{n_1=1}^{\lfloor (P+1)/2 \rfloor} \sum_{n_2=1}^{\lfloor P/2 \rfloor} 4h[2n_1 - 1, 2n_2]\{\cos(2n_1 - 1)\omega_1 \cos(2n_2)\omega_2 \\ &\quad - \cos(2n_2)\omega_1 \cos(2n_1 - 1)\omega_2\}. \end{aligned} \tag{4.25}$$

Thus, the number of independent filter coefficients N_i is the same as that shown in Eq. (4.15).

We can directly design a fan filter using our frequency sampling method. We place N_i samples in a triangular area within the passband, as shown in Fig. 4.21, since this is the only independent part of the amplitude response.

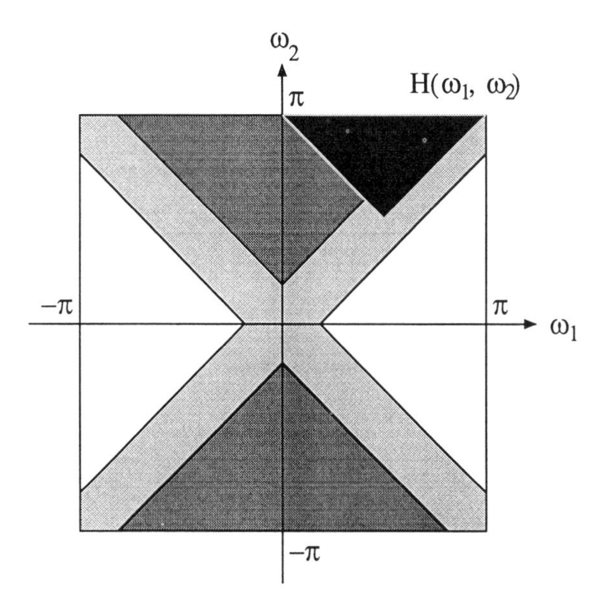

Figure 4.21. Amplitude response of a fan filter. The darkly shaded region is the only independent part, because of eightfold symmetry and the half-band nature of the filter.

The process of generating the samples depends on the fact that the cross-section of $A(\omega_1, \omega_2)$ along the diagonal ($\omega_1 = \pi - \omega_2$) looks like a 1-D half-band lowpass response. This is similar to the process of generating samples for a diamond filter in Section 4.6. If H_k is a sample for the corresponding diamond filter at the location $(\omega_{1k}, \omega_{2k})$, then H_k is the sample to be used at the location $(\omega_{1k}, \pi - \omega_{2k})$ for designing the fan filter. Once the N_i samples are generated, they are used to solve for the filter coefficients in Eq. (4.25).

In the following example, we design a fan filter using our method, and then compare the results with those obtained by three other design methods.

Example 4.4

Consider the design of a fan filter with the following specifications:

Support size = 17×17,
Passband edge $\omega_p = 0.43\pi$, Stopband edge $\omega_s = 0.57\pi$.

Only 20 of the 289 filter coefficients are independent. The amplitude response of the fan filter designed is shown in Fig. 4.22(a). The sample locations used are shown superimposed on the contour plot in Fig. 4.22(b). These figures show that a very good passband shape is obtained along with extremely low peak ripples.

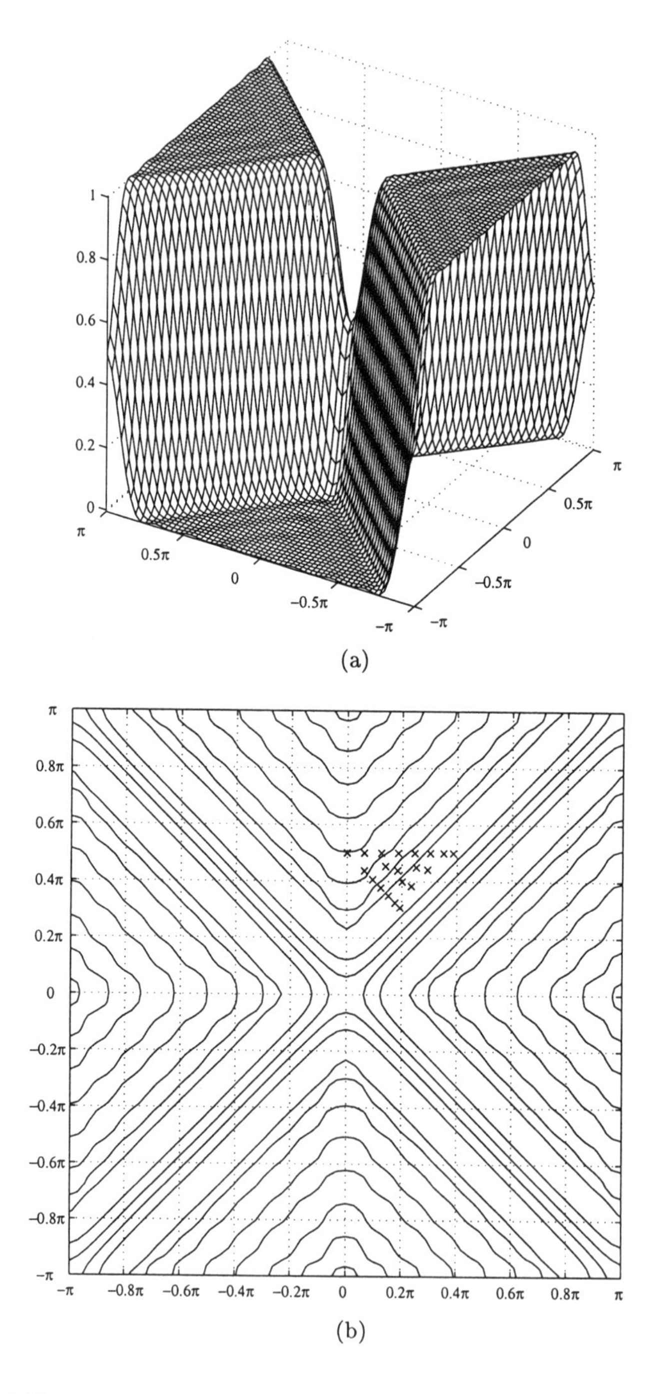

Figure 4.22. Fan filter of size 17×17 designed by the NDFT method. (a) Amplitude response. (b) Contour plot. The sample locations are denoted by "$\times$".

We compare the above filter to three other filters designed with the same specifications, using the methods:

(1) Frequency transformation.

(2) Ansari's method [Ansari and Reddy, 1987].

(3) Iterative l_p optimization technique proposed by Lodge and Fahmy [Lodge and Fahmy, 1980].

The peak ripples obtained are shown in Table 4.5. The amplitude response and contour plots of the filters designed by Methods 1 and 2 are shown in Figs. 4.23 and 4.24, respectively. The NDFT method gives the lowest peak ripple. Besides, the shapes of the contours obtained are much better than with the other methods.

In the frequency transformation method, we design a fan filter by applying the following 3×3 transformation to a 1-D half-band lowpass filter:

$$\cos \omega = T(\omega_1, \omega_2) = \frac{1}{2} \cos \omega_1 - \frac{1}{2} \cos \omega_2. \tag{4.26}$$

The contours tend to be circular, which is undesirable.

In Ansari's method, a 1-D half-band lowpass filter of length 17 is used as a prototype. This filter is used to obtain a 2-D filter that has a passband in the second and fourth quadrants, and a stopband in the first and third quadrants of the frequency plane. This 2-D filter is then transformed to the desired fan filter by a 45-degree rotation in frequency. Though the shape of the passband is better than in frequency transformation, the peak ripple is also larger.

For Method 3, we refer to an example presented by Aly and Fahmy, where a fan filter was designed with the same specifications as in Example 4.4 [Aly and Fahmy, 1981, Ex. 1, p. 800]. The contour plot of this 17×17 fan filter was reported by them [Aly and Fahmy, 1981, Fig. 7, p. 801]. The contours tend to become circular near $(\pi, 0)$ and $(0, \pi)$.

Table 4.5. Performance comparison for fan filter design.

Method	δ_p	δ_s
NDFT	0.0051	0.0051
Frequency Transformation	0.0238	0.0238
Ansari	0.0365	0.0352
l_p Optimization	0.0223	0.0223

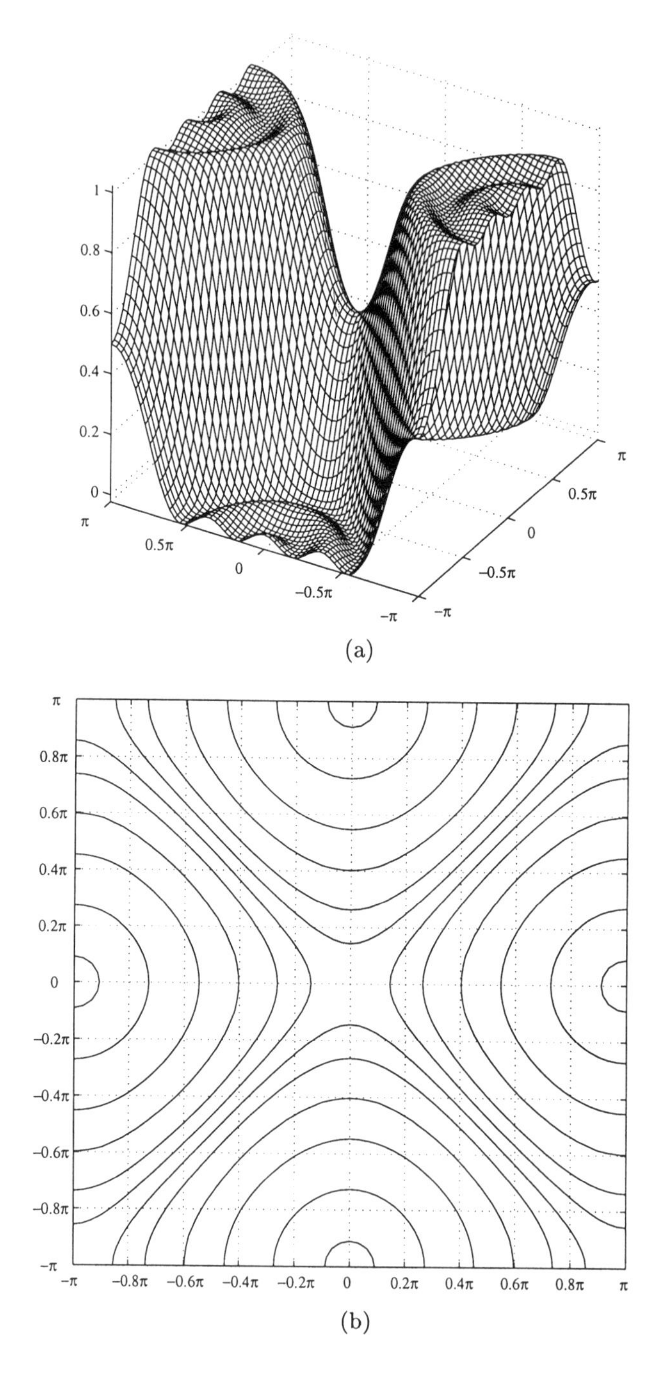

Figure 4.23. Fan filter of size 17 × 17 designed by the McClellan frequency transformation method. (a) Amplitude response. (b) Contour plot.

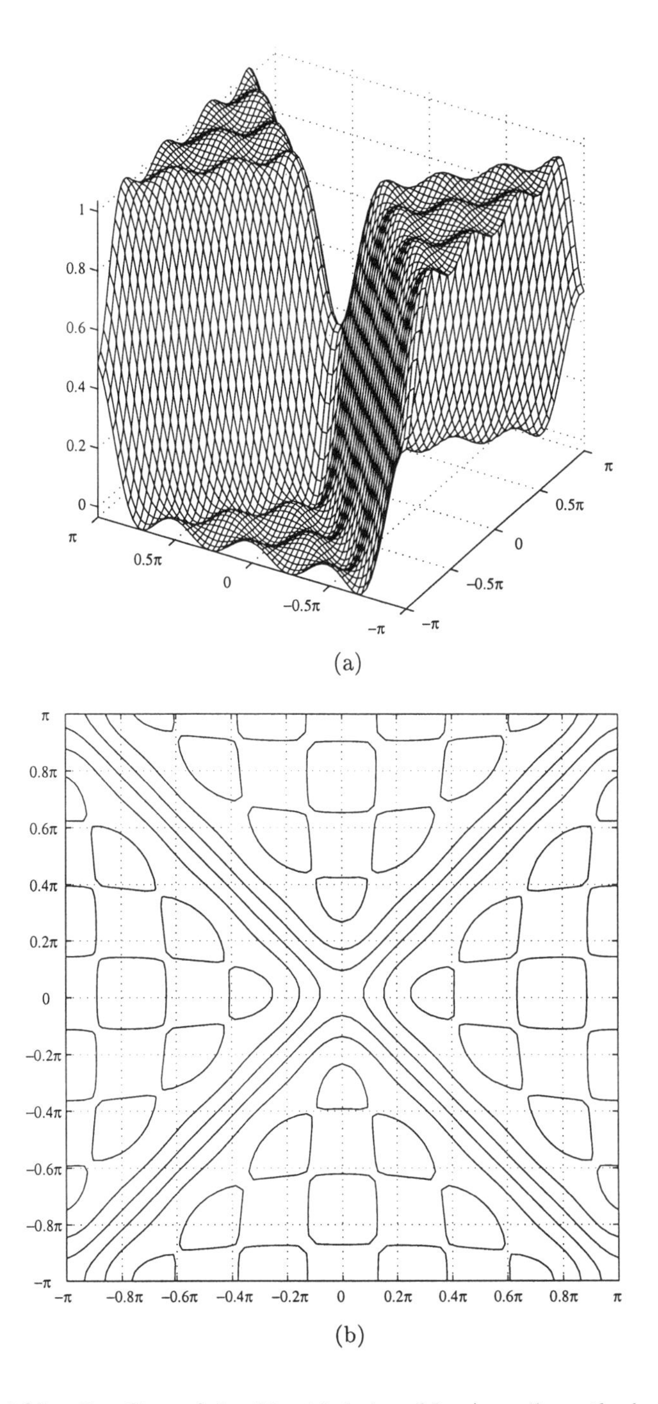

Figure 4.24. Fan filter of size 17×17 designed by Ansari's method. (a) Amplitude response. (b) Contour plot.

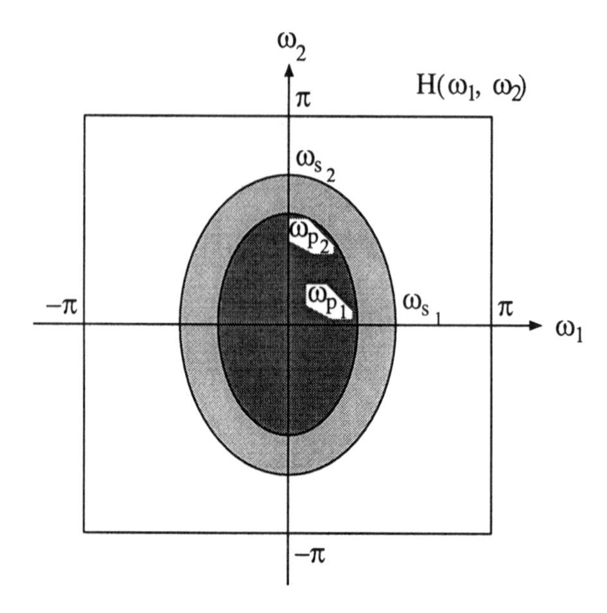

Figure 4.25. Amplitude-response specification of an elliptically-shaped lowpass filter. Darkly shaded region: passband, lightly shaded region: transition band, unshaded region: stopband.

4.7 ELLIPTICALLY-SHAPED LOWPASS FILTER DESIGN

The amplitude-response specification of an elliptically-shaped 2-D lowpass filter is shown in Fig. 4.25. Let us consider the design of this filter using the proposed approach. Suppose that the filter is of size $N_1 \times N_2$, and has passband and stopband edges described by the following elliptical contours:

$$\text{Passband contour:} \quad \left(\frac{\omega_1}{\omega_{p_1}}\right)^2 + \left(\frac{\omega_2}{\omega_{p_2}}\right)^2 = 1, \tag{4.27}$$

$$\text{Stopband contour:} \quad \left(\frac{\omega_1}{\omega_{s_1}}\right)^2 + \left(\frac{\omega_2}{\omega_{s_2}}\right)^2 = 1. \tag{4.28}$$

Further, let ω_{p_1} and ω_{s_1} represent the lengths of the minor axes of these ellipses, so that $\omega_{p_1} < \omega_{p_2}$ and $\omega_{s_1} < \omega_{s_2}$. From Fig. 4.25, it is evident that the amplitude response has a fourfold symmetry as shown in Eq. (4.4). Thus, $A(\omega_1, \omega_2)$ can be expressed in the form

$$A(\omega_1, \omega_2) = h[0,0] + \sum_{n_1=1}^{(N_1-1)/2} 2h[n_1, 0] \cos \omega_1 n_1 + \sum_{n_2=1}^{(N_2-1)/2} 2h[0, n_2] \cos \omega_2 n_2$$

$$+\sum_{n_1=1}^{(N_1-1)/2}\sum_{n_2=1}^{(N_2-1)/2} 4h[n_1,n_2]\cos\omega_1 n_1 \cos\omega_2 n_2. \tag{4.29}$$

Therefore, the number of independent filter coefficients is given by

$$N_i = \frac{(N_1+1)(N_2+1)}{4}. \tag{4.30}$$

To solve for these coefficients, we place N_i samples of $A(\omega_1,\omega_2)$ on elliptical contours in the first quadrant of the (ω_1,ω_2) plane. In doing so, we utilize the fact that the cross-section of the 2-D amplitude response along the minor axis of the ellipses (the ω_1 axis in this case) looks like a 1-D lowpass filter response. This 1-D response is represented by analytic functions, as shown in Chapter 3 (Section 3.3.1). Then the $(N_1+1)/2$ samples along the minor axis are placed at the extrema of this 1-D response. The other samples are placed on ellipses passing through these extrema. The ellipses within the passband are chosen to have the same eccentricity as the passband contour in Eq. (4.27). Similarly, those in the stopband have the same eccentricity as the stopband contour in Eq. (4.28). All samples on a particular ellipse have the same value and are spaced at equal angles from each other. The total number of samples must equal N_i. The distribution of samples among the different contours is illustrated in the following design example.

Example 4.5

Consider the design of an elliptically-shaped filter with a square support and the following specifications:

Support size $= 9 \times 9$,
Along minor axis: $\omega_{p_1} = 0.2\pi$, $\omega_{s_1} = 0.4\pi$,
Along major axis: $\omega_{p_2} = 0.4\pi$, $\omega_{s_2} = 0.6\pi$.

The 25 samples required are placed on elliptical contours, as shown in Fig. 4.26(b) (superimposed on the contour plot of the resulting filter). The amplitude response is shown in Fig. 4.26(a). The contours corresponding to the specified band-edges and the ones actually obtained are plotted in Fig. 4.26(c). Evidently, the specifications are met very closely.

We now compare the above filter to those obtained by using the frequency transformation method [Mersereau et al., 1976], with the same specifications. In this method, a frequency transformation must be designed so as to give the desired elliptical band-edges. This is then used to transform a 1-D filter to the desired 2-D filter. For the required 9×9 filter, we use a transformation of size 3×3 and a 1-D filter of length 9. We design the transformation by following the constrained approximation approach with a least squares formulation [Mersereau et al., 1976]. Two cases are of interest:

(1) The passband edge ω_{p0} of the 1-D filter maps to the passband contour in Eq. (4.27), with no constraints on the stopband.

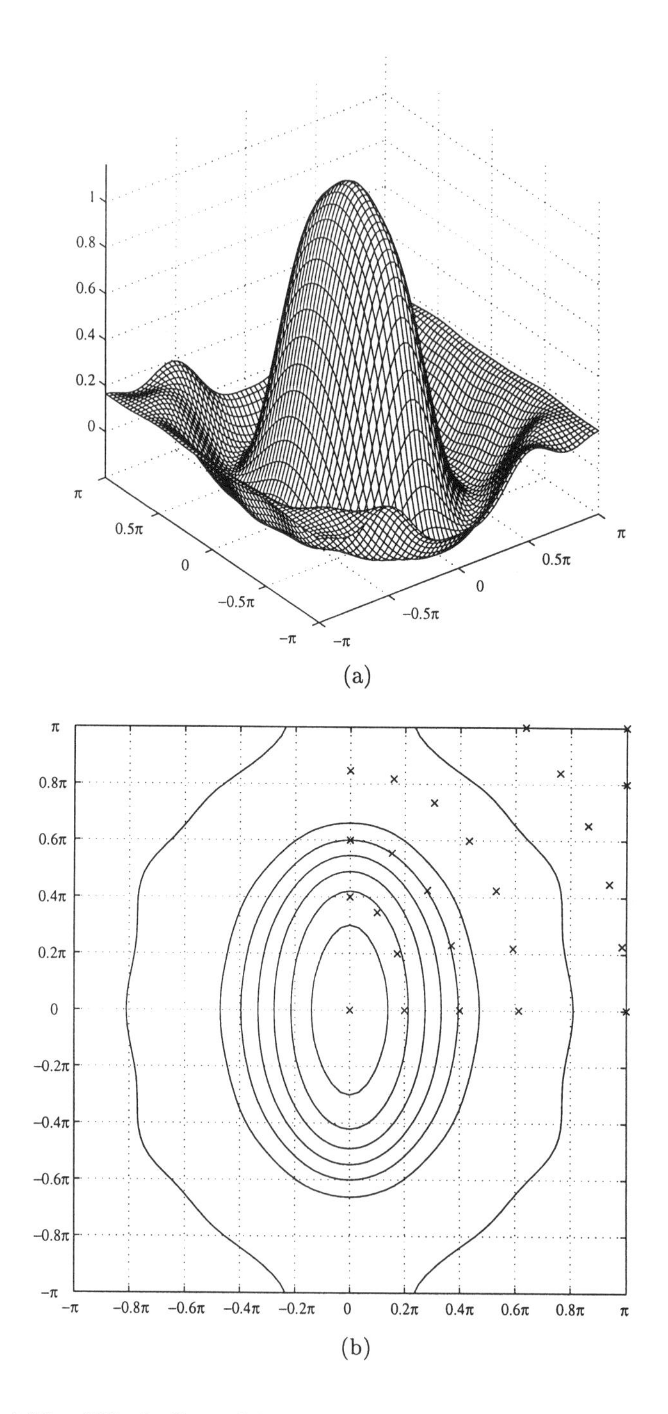

Figure 4.26. Elliptic filter of size 9×9 designed by the NDFT method. (a) Amplitude response. (b) Contour plot. The sample locations are denoted by "$\times$".

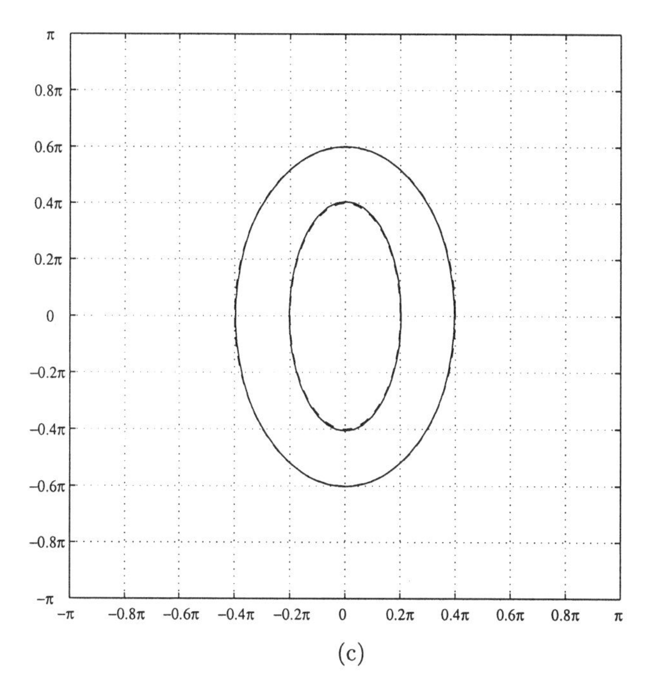

Figure 4.26. (continued) (c) Contour plot showing the band-edges actually obtained (solid line) and the specified band-edges (dotted line). Since these almost coincide, the filter attains the specifications very closely.

(2) Both the passband and stopband edges ω_{p_0} and ω_{s_0} of the 1-D filter map to the 2-D contours given in Eqs. (4.27) and (4.28), respectively.

For clarity, we briefly outline the method used for Case 1.
Let the transformation have the form

$$\begin{aligned}\cos\omega &= T(\omega_1, \omega_2) \\ &= t[0,0] + t[1,0]\cos\omega_1 + t[0,1]\cos\omega_2 + t[1,1]\cos\omega_1\cos\omega_2. \quad (4.31)\end{aligned}$$

Since both the 1-D and 2-D filters are lowpass, we require that the 1-D origin map to the 2-D origin. This gives

$$t[0,0] + t[1,0] + t[0,1] + t[1,1] = 1. \quad (4.32)$$

Thus, there are three independent parameters to be solved for. Along the 2-D passband edge, we have

$$\omega_2 = \omega_{p_2}\sqrt{1 - \left(\frac{\omega_1}{\omega_{p_1}}\right)^2}. \quad (4.33)$$

We formulate the problem as a linear approximation problem as follows. If the desired mapping was exactly obtained, then the value of $T(\omega_1, \omega_2)$ would be constant as we traversed the elliptical passband contour. Thus, we would get

$$\begin{aligned}\cos \omega_{p_0} &= t[0,0] + t[1,0]\cos(\omega_1) + t[0,1]\cos\left\{\omega_{p_2}\sqrt{1-\left(\frac{\omega_1}{\omega_{p_1}}\right)^2}\right\} \\ &\quad + t[1,1]\cos\omega_1 \cos\left\{\omega_{p_2}\sqrt{1-\left(\frac{\omega_1}{\omega_{p_1}}\right)^2}\right\}. \end{aligned} \tag{4.34}$$

However, in practice, this equality is satisfied only approximately since the mapping is not exact. We can then define an error function as

$$\begin{aligned}E(\omega_1) &= \cos \omega_{p_0} - t[0,0] - t[1,0]\cos(\omega_1) - t[0,1]\cos\left\{\omega_{p_2}\sqrt{1-\left(\frac{\omega_1}{\omega_{p_1}}\right)^2}\right\} \\ &\quad -(1 - t[0,0] - t[0,1] - t[1,0])\cos\omega_1 \cos\left\{\omega_{p_2}\sqrt{1-\left(\frac{\omega_1}{\omega_{p_1}}\right)^2}\right\}. \end{aligned} \tag{4.35}$$

Since this is a linear function of the unknown parameters, we can minimize the squared error using a least square approach to obtain the transformation parameters.

In Case 2, the least square fit is done for both the passband and stopband edges. The resulting 9×9 filters for Cases 1 and 2 are shown in Figs. 4.27 and 4.28, respectively. Figs. 4.27(c) and 4.28(c) show the specified band edges and the ones actually obtained for the two cases. Since the transformation has a low order, it is not possible to get exactly the desired band-edges, even in Case 2.

The NDFT method provides better control over the shapes of the band edges. The peak ripples obtained for the filters are shown in Table 4.6. The peak stopband ripple for Case 1 is quite large since the stopband is not well-constrained.

Table 4.6. Performance comparison for elliptic filter design.

Method	δ_p	δ_s
NDFT	0.1658	0.1971
Frequency Transformation 1	0.1151	0.3000
Frequency Transformation 2	0.1539	0.1565

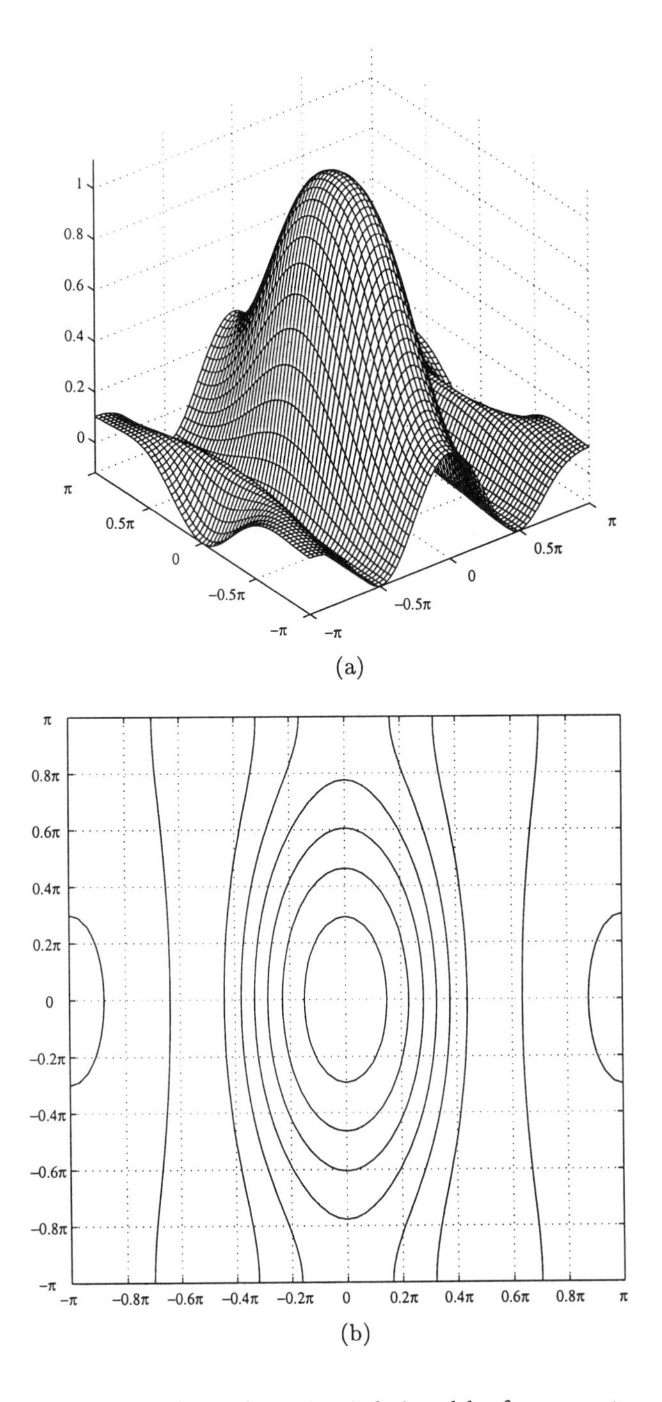

Figure 4.27. Elliptic filter of size 9×9 designed by frequency transformation, Method 1. (a) Amplitude response. (b) Contour plot.

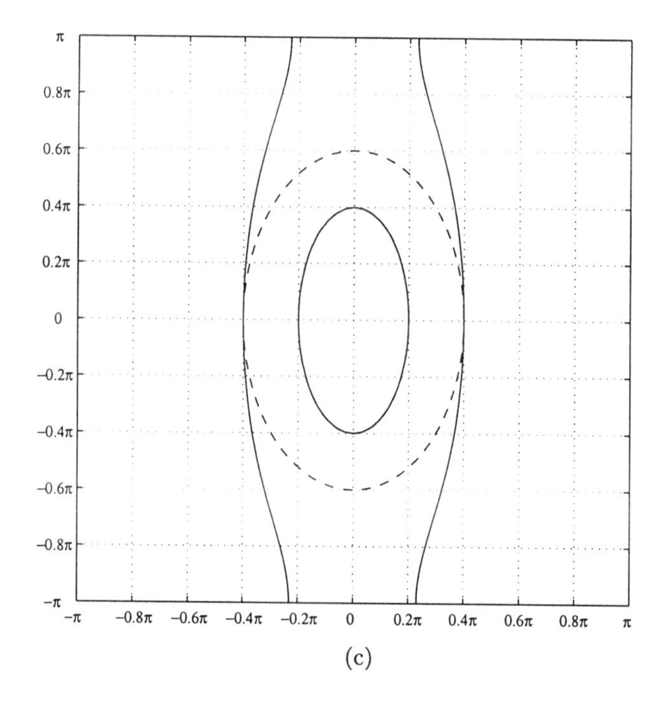

Figure 4.27. (continued) (c) Contour plot showing the band-edges actually obtained (solid line) and the specified band-edges (dotted line). Though the passband specifications are met, the stopband edge is not well constrained.

This can be seen by comparing the amplitude responses in Figs. 4.27(a) and 4.28(a).

Next, we give an example to demonstrate the flexibility of the NDFT method. This involves the design of a filter with a rectangular support. This example is interesting since the frequency transformation method cannot design a filter with the given specifications. The latter method can design filters with rectangular support only if the transformation itself is made rectangular, with size $(2P+1)\times(2Q+1)$. The resulting 2-D filter can have only a limited set of sizes, $(2PM+1)\times(2QM+1)$, where $2M+1$ is the length of the 1-D filter used.

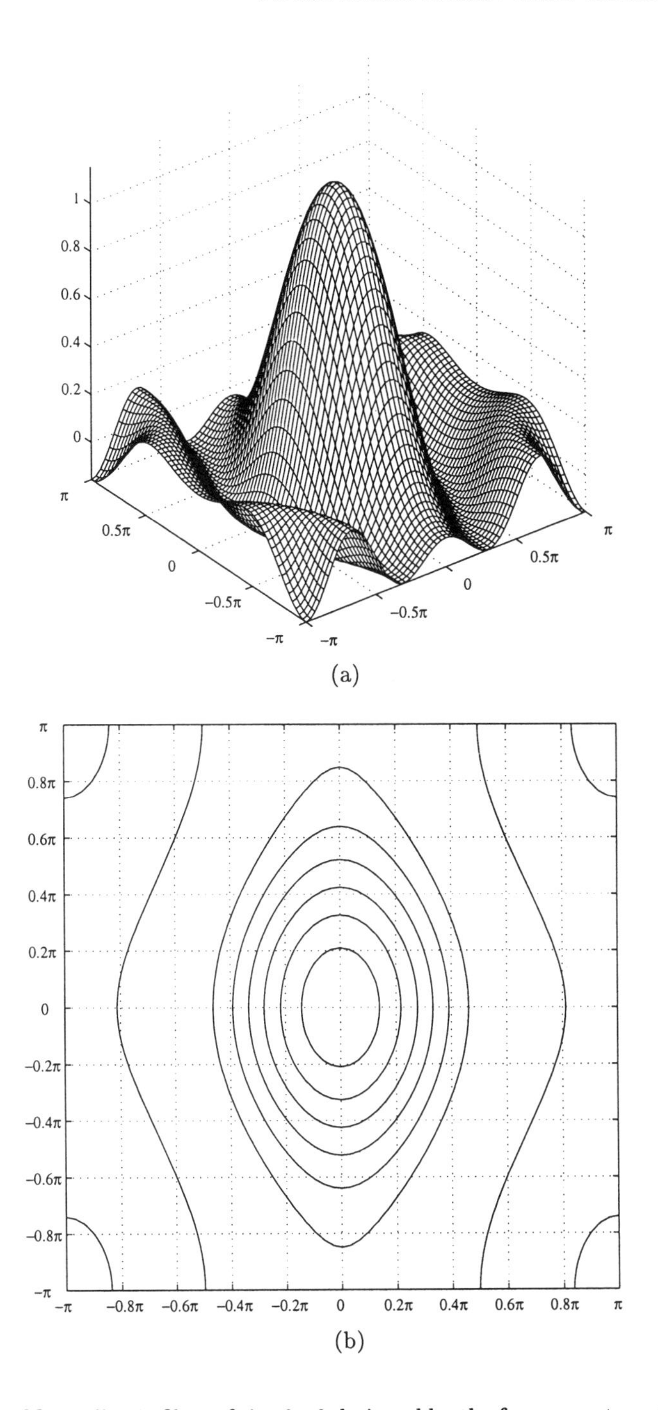

Figure 4.28. Elliptic filter of size 9×9 designed by the frequency transformation, Method 2. (a) Amplitude response. (b) Contour plot.

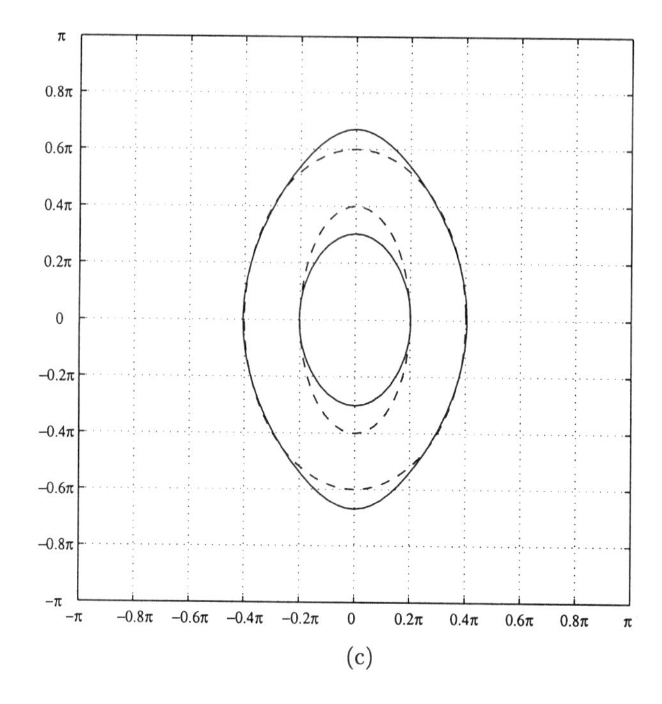

Figure 4.28. (continued) (c) Contour plot showing the band-edges actually obtained (solid line) and the specified band-edges (dotted line). Note the differences along the major axis.

Example 4.6

Consider the design of an elliptically-shaped filter with a rectangular support and the following specifications:

Support size $= 9 \times 7$,
Along minor axis: $\omega_{p_1} = 0.2\pi$, $\omega_{s_1} = 0.4\pi$,
Along major axis: $\omega_{p_2} = 0.4\pi$, $\omega_{s_2} = 0.6\pi$.

The 20 samples required are placed on elliptical contours as shown in Fig. 4.29(b). The resulting amplitude response is shown in Fig. 4.29(a). By taking a look at the contours in Fig. 4.29(c), we can see that the band-edge specifications are met quite closely. The peak ripples obtained are:

$\delta_p = 0.1786, \qquad \delta_s = 0.1774.$

As compared to the filter designed by the NDFT in Example 4.5, this filter has slightly higher δ_p and lower δ_s. Since the support size is also smaller, the performance is quite good.

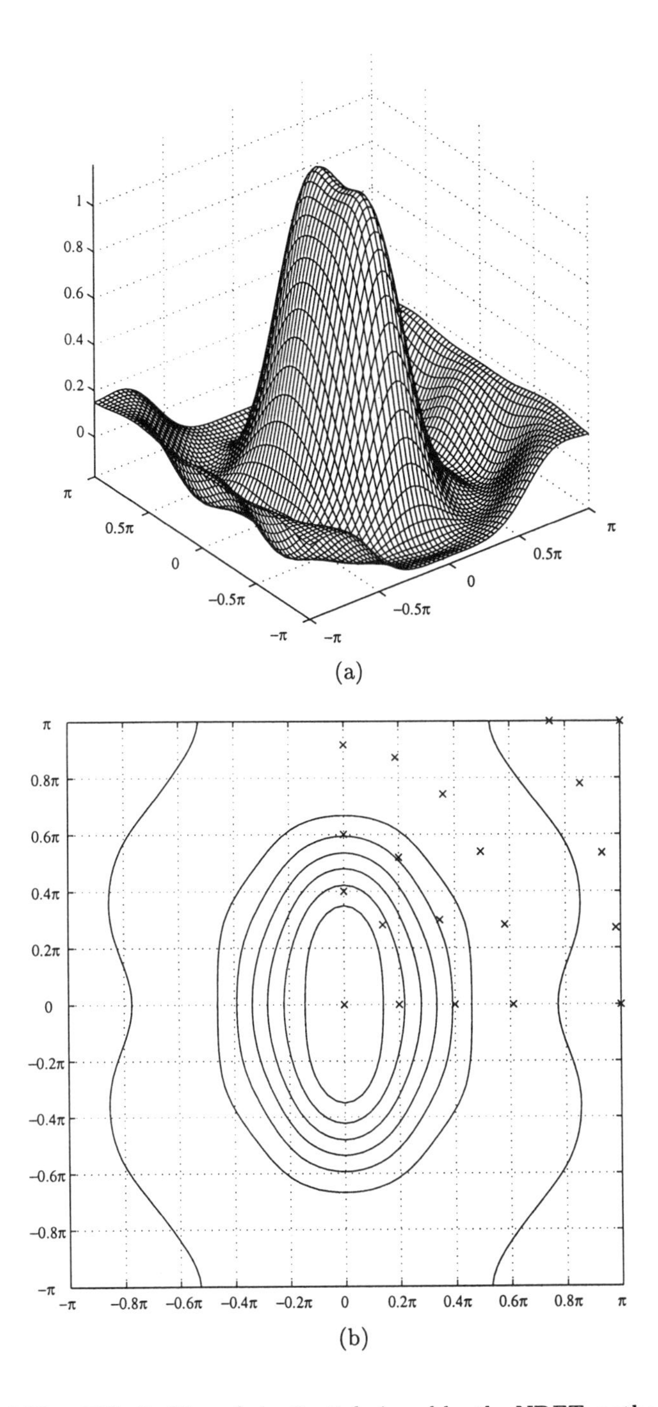

Figure 4.29. Elliptic filter of size 9×7 designed by the NDFT method. (a) Amplitude response. (b) Contour plot. The sample locations are denoted by "$\times$".

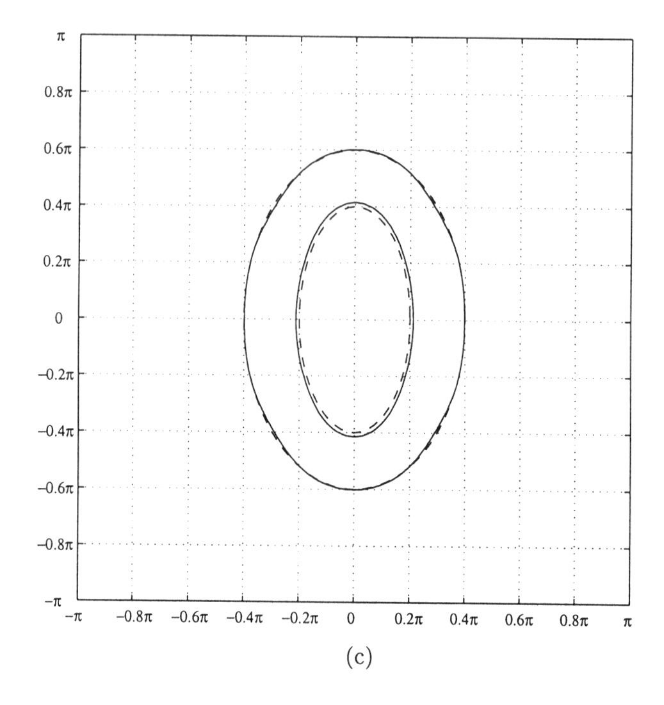

Figure 4.29. (continued) (c) Contour plot showing the band-edges actually obtained (solid line) and the specified band-edges (dotted line). Since these are nearly the same, the filter attains the specifications closely.

4.8 APPLICATIONS OF 2-D FILTERS

Two-dimensional filters have numerous applications in digital signal processing. They are used in the area of image processing for enhancement of low-quality images such as X-ray photographs and satellite pictures, and for conversion between different sampling structures [Siohan, 1991] and video formats. They find applications as prefilters in various schemes for image coding, and in filter banks of subband analysis-synthesis systems. They are also used for processing multidimensional data obtained from geophysical exploration, seismology, sonar, radar and radio astronomy. The major aspects involved in the real-time operation of 2-D filters are the large amount of computational effort and memory needed to implement them. Recent advances in VLSI technology have made real-time 2-D filtering practically feasible.

We will now consider examples involving the application of square and diamond filters. These filters are widely used as *prefilters* prior to downsampling, and as *postfilters* for interpolating zero-valued samples after upsampling of images. Square-shaped filters are used in schemes for rectangular downsampling, whereas diamond-shaped filters are used for quincunx downsampling.

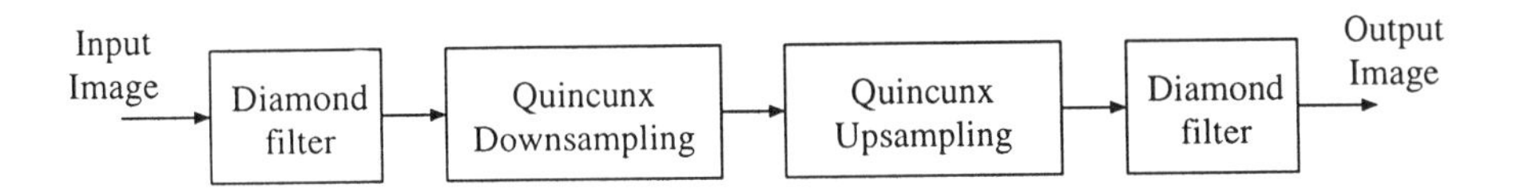

Figure 4.30. A quincunx downsampling scheme with diamond filters as prefilters and postfilters.

Example 4.7

We consider as an example the application for diamond filters as prefilters and postfilters in a quincunx downsampling scheme, shown in Fig. 4.30. Such schemes are used to reduce the data rate for digital transmission of HDTV signals. Quincunx downsampling is preferred to orthogonal downsampling because the former does not limit resolution in the horizontal and vertical directions—the human visual system is more sensitive along these directions. We compare the performances of two diamond filters in Example 4.3, designed by the NDFT method and Bamberger-Smith's method. The original Lena image is shown in Fig. 4.32. Fig. 4.33 shows the output images. With the input image $x[n_1, n_2]$ as reference, we compute the peak signal-to-noise ratio (PSNR) for each output image $y[n_1, n_2]$, as defined by [Pratt, 1991]

$$\text{PSNR} = 10 \log_{10} \frac{[\max\{x[n_1, n_2]\}]^2}{\sum_{n_1=0}^{N-1} \sum_{n_2=0}^{N-1} |x[n_1, n_2] - y[n_1, n_2]|^2}, \tag{4.36}$$

where the image size is $N \times N$, and $\max\{x[n_1, n_2]\} = 255$ for an image with 8 bits/pixel. A comparison of the PSNR values is given in Table 4.7. The image produced by using Bamberger-Smith's diamond filter appears to have lower brightness and contrast, and thus, has a lower PSNR.

Table 4.7. PSNR for diamond filters in quincunx downsampling scheme.

Filter design method	PSNR (dB)
NDFT	35.12
Bamberger-Smith	20.70

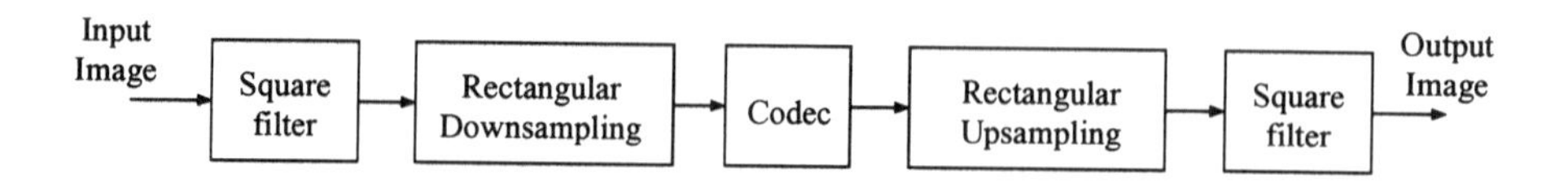

Figure 4.31. A rectangular downsampling scheme with square filters as prefilters and postfilters.

Example 4.8

In this example, we evaluate the performance of two square filters designed in Example 4.1 by the NDFT and the separable design methods. These filters are applied in a typical scheme for rectangular downsampling shown in Fig. 4.31. We used the JPEG codec [Leger et al., 1991] to observe the effect of coding the smaller, downsampled image. Fig. 4.34 shows the output images, without and with the codec in the system. The overall bit rate with the codec is 0.5 bits/pixel. This includes the 4:1 reduction due to downsampling. For reference, the input image was also coded using JPEG only (without downsampling) at 0.5 bits/pixel, as shown in Fig. 4.35. Although this image is sharper, it exhibits strong block artifacts, visible in the plain regions. Such artifacts are not present in the images produced by the downsampling scheme, due to the smoothing effect of the filters. The PSNR values are given in Table 4.8. The square filter designed by the NDFT method performs better than the separable square filter in both cases.

Table 4.8. PSNR for square filters in rectangular downsampling scheme.

Filter design method	PSNR (dB)	
	Without codec	With codec
NDFT	29.68	29.05
Separable	24.99	24.83

4.9 SUMMARY

In this chapter, we proposed a method for designing 2-D FIR filters, based on the concept of the 2-D NDFT. This method utilizes the freedom of locating samples nonuniformly in the frequency plane, to produce nonseparable 2-D filters with good passband shapes and low peak ripples. We demonstrated the design of 2-D filters with a variety of passband shapes, and compared our results with other existing design methods. These comparisons have established the effectiveness of the proposed method. Some filters designed were tested by applying them in schemes for downsampling of images. Considering the general problem of nonuniform frequency sampling, we laid down some specific guidelines regarding the choice of the sample values and locations, which were lacking in earlier techniques. These guidelines can serve as a basis for the design of more complex 2-D filters. Finally, the proposed method can be extended to multidimensions for designing 3-D filters which are used for filtering video signals.

Figure 4.32. Original Lena image.

(a)

(b)

Figure 4.33. Images produced by quincunx downsampling scheme, with diamond filters designed by: (a) NDFT method, (b) Bamberger-Smith's method.

(a)

(b)

Figure 4.34. Images produced by rectangular downsampling scheme, with square filters designed by: (a) NDFT method, (b) Separable design.

(c)

(d)

Figure 4.34. (continued) Images produced by rectangular downsampling scheme (using codec on smaller images), with square filters designed by: (c) NDFT method, (d) Separable design.

Figure 4.35. Lena image coded by JPEG at 0.5 bits/pixel.

5 ANTENNA PATTERN SYNTHESIS WITH PRESCRIBED NULLS

5.1 INTRODUCTION

In this chapter, we present an application of the NDFT in antenna pattern synthesis. An array antenna is often subjected to interference signals from certain specific directions. These signals can be caused by interference between wireless communication lines. In such situations, adaptive phased array radar (APAR) spectral estimation techniques are used to determine the angular location of the interfering signals. Using this information, the antenna pattern is designed such that the undesired signals are suppressed by placing nulls on the pattern at the specified angles. Techniques for pattern synthesis with prescribed nulls have become increasingly important because of the rising pollution of the electromagnetic environment.

The NDFT approach is ideally suited to this problem since we can achieve pattern nulls at *exactly the desired angles* by placing zero-valued samples on the antenna pattern at these angles. The proposed synthesis technique, based on nonuniform sampling in the angular domain, can produce patterns with *deep nulls* without decreasing the array directivity in the main beam direction.

The outline of this chapter is as follows. In Section 5.2, we discuss some existing techniques for synthesizing array patterns with nulls. The proposed NDFT method for null synthesis is presented in Section 5.3. We consider linear arrays of equispaced elements, with sinc patterns and Chebyshev patterns. In

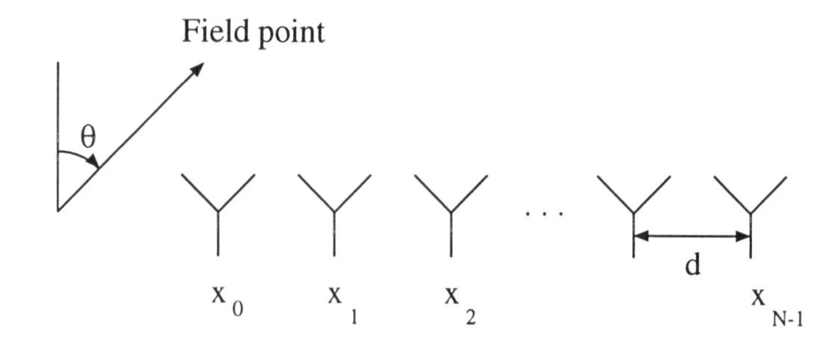

Figure 5.1. Linear array of N equispaced elements.

Section 5.4, we present design examples to illustrate the effectiveness of the NDFT method, and compare them with patterns designed by other methods.

5.2 EXISTING METHODS FOR NULL SYNTHESIS

Let us consider a linear array of N isotropic, equispaced elements. Its pattern is defined as

$$p(u) = \sum_{n=0}^{N-1} a_n e^{jnkdu}, \tag{5.1}$$

where a_n denotes the complex excitation or weight of the nth element, k is the wavenumber (defined as $2\pi/\lambda$, for wavelength λ), d is the interelement spacing, and u is the angular parameter

$$u = \sin\theta, \tag{5.2}$$

θ being the angle from the broadside (i.e., the normal to the array), as illustrated in Fig. 5.1. Examining Eq. (5.1), we note that the array pattern $p(u)$ corresponds to the discrete-time Fourier transform of the sequence of weights, a_n. Therefore, there is an *analogy* between the pattern of an array and the frequency response of a filter.

Control over the nulls in an antenna pattern can be achieved in various ways. Depending upon the complexity of electronic control, various structures are obtained. These include control of the amplitude and phase of each element, control over the phase only, control of a select subset of elements, or control at the subarray level [Steykal et al., 1986].

One family of existing techniques is based on constrained least square approximation. Steykal et al. have presented an overview of null synthesis methods employing this approach [Steykal et al., 1986]. The constrained pattern is designed as an approximation to the original pattern by minimizing the mean square difference between the original and the constrained patterns, subject to a set of null constraints [Steykal, 1982].

A single pattern null at the proper angle can suppress a narrow-band interference source. However, a wide-band jammer appears to be smeared out over an angular sector of the pattern because of the frequency dependence of the antenna, and therefore, requires nulling over an entire sector. Steykal has considered two methods to achieve a *broad pattern null*—imposing nulls on the pattern at closely spaced angles over the sector, and imposing nulls on the pattern and its derivatives at the center of the angular sector [Steykal, 1983]. These methods are referred to as multiple nulling and higher order nulling, respectively. Results show that the multiple nulling technique is more effective for wide-band sidelobe suppression [Steykal, 1983]. For the case when M single nulls are to be placed on the array $p_0(u)$ with N elements and half-wavelength spacing ($d = \lambda/2$), the constrained pattern obtained by the least square technique may be written as

$$p_c(u) = p_0(u) - \sum_{m=1}^{M} \alpha_m q_m(u), \tag{5.3}$$

where the cancellation beams $q_m(u)$ are sinc-like functions given by

$$q_m(u) = \frac{1}{N} \frac{sin[\pi N(u - u_m)/2]}{sin[\pi(u - u_m)/2]}, \tag{5.4}$$

and u_m denotes the location of the mth imposed null [Steykal, 1983]. The beam amplitudes α_m are determined by solving the M equations for the nulls,

$$\begin{bmatrix} p_0(u_1) \\ p_0(u_2) \\ \vdots \\ p_0(u_M) \end{bmatrix} = \begin{bmatrix} q_1(u_1) & q_2(u_1) & \cdots & q_M(u_1) \\ q_1(u_2) & q_2(u_2) & \cdots & q_M(u_2) \\ \vdots & \vdots & \ddots & \vdots \\ q_1(u_M) & q_2(u_M) & \cdots & q_M(u_M) \end{bmatrix} \begin{bmatrix} \alpha_1 \\ \alpha_2 \\ \vdots \\ \alpha_M \end{bmatrix}. \tag{5.5}$$

This null synthesis method is computationally *efficient*, and gives good results. However, the directivity in the direction of the main beam is reduced slightly in most cases. Besides, the closed solution discussed here holds only for half-wavelength interelement spacing.

Another technique for null synthesis was proposed by Carroll and Kumar [Carroll and Kumar, 1989]. This is an application of the error reduction algorithm, which uses the DFT iteratively to match constraints in the time and frequency domains. In this method, the N weights of the original pattern are padded with zeros and a P-point DFT is applied to obtain samples of the pattern at P uniformly spaced angles. Nulls are then imposed by inserting zero-values for the pattern samples closest to the desired null angles. The resulting pattern samples are transformed back to the weight domain by applying a P-point inverse DFT. The weight vector thus obtained has P elements, and is truncated to obtain the N weights of the constrained pattern. This process is then repeated in successive iterations. Note that, to obtain a good resolution in the angular domain, P must be chosen to be much larger than N (e.g. $P = 512$ or 1024, an FFT can be used). Higher values of P can give deeper nulls, but also result in slower convergence. Another point to be noted is that, the truncation of the weight vector is analogous to rectangular windowing of a filter impulse response. Therefore, the constrained pattern is convolved with the Fourier transform of a rectangular window (a sinc function). This causes a smearing of the pattern and also disturbs the location of the nulls. The performance of this null synthesis technique is, therefore, undermined by two factors. Firstly, the nulls *cannot be placed* at exactly the desired locations because of the fixed uniform sampling of the pattern given by the DFT. Secondly, the nulls imposed in the angular domain *are not preserved* since they are affected by the following truncation in the weight domain. Consequently, this algorithm gives nulls of poor quality, even after a *large* number of iterations (e.g., 1000). This is illustrated later in a design example in Section 5.4 (Example 5.2).

5.3 PROPOSED NULL SYNTHESIS METHOD

We have utilized the concept of the NDFT to develop a method of synthesizing antenna patterns with nulls at a given set of angles. Since the weights of a linear N-element array form a sequence of length N, its NDFT corresponds to samples of the array pattern evaluated at N distinct, nonuniformly spaced angles. From Eq. (5.1), the NDFT is given by

$$p(u_k) = \sum_{n=0}^{N-1} a_n e^{jnkdu_k}, \qquad k = 0, 1, \ldots, N-1, \tag{5.6}$$

where

$$u_k = \sin\theta_k, \qquad k = 0, 1, \ldots, N-1, \tag{5.7}$$

and $\theta_0, \theta_1, \ldots, \theta_{N-1}$ are any N distinct angles.

We have considered two types of patterns for null synthesis—sinc patterns and Chebyshev patterns. These patterns are *complementary* in a sense, because sinc patterns have sidelobes of constant width and varying height, whereas Chebyshev patterns have sidelobes of constant height and varying width.

Suppose we require M nulls to be placed on the pattern $p(u)$ at $u = w_1, w_2, \ldots, w_M$. The proposed design procedure can be described by the following steps:

1. Represent the original pattern $p(u)$ by means of its NDFT $p(u_k)$ computed at N points, $u_0, u_1, \ldots, u_{N-1}$, which include the extrema of the pattern.

2. Impose M nulls on the pattern by changing the M samples of $p(u)$ closest to the given nulls as follows. Let the samples at $u = u_{i_1}, u_{i_2}, \ldots, u_{i_M}$ be the closest in angular location to the given nulls, $w_1, w_2, \ldots, w_M$, where $i_1, i_2, \ldots, i_M$ are integers in the range $(0, N-1)$. Then, we replace the M samples $p(u_{i_1}), p(u_{i_2}), \ldots,$ $p(ui_M)$ by zeros located at $u = w_1, w_2, \ldots, w_M$. The remaining $N - M$ samples are not changed.

3. Apply the inverse NDFT to these N samples to obtain the weights of the constrained pattern with the prescribed nulls.

4. If needed, another iteration is performed as follows. The new array pattern is first computed at a densely spaced set of angles by applying an FFT to the zero-padded weight vector. Then, the angles of the $N - M$ non-zero samples are moved to the nearest pattern extrema, keeping the sample values unchanged. The M null samples are left undisturbed. An inverse NDFT is applied to these N samples to obtain the new array weights.

5. Go to Step 4, if required.

As we shall see later from the design examples in Section 5.4, this design procedure yields good results in just two or three iterations. We shall now take a look at sinc patterns and Chebyshev patterns, and provide more details regarding the placement of samples used in Step 1.

5.3.1 *Sinc Pattern*

Consider a linear array of N equispaced elements with a uniform current distribution and linear phase progression [Collin and Zucker, 1969]. Its excitation coefficients a_n all have the same absolute value and a constant phase difference δ between one excitation coefficient and the next, as given by

$$a_n = \frac{1}{N} e^{jn\delta}, \qquad n = 0, 1, \ldots, N-1. \tag{5.8}$$

Applying Eq. (5.1), the array pattern can be expressed as

$$p(u) = \frac{1}{N} \sum_{n=0}^{N-1} e^{jn(\delta + kdu)}. \tag{5.9}$$

Thus, the pattern can be written as a function of the variable,

$$\nu = \delta + kdu. \tag{5.10}$$

On doing this, we obtain

$$p(\nu) = \frac{1}{N} \sum_{n=0}^{N-1} e^{jn\nu}. \tag{5.11}$$

This is the expression for the sum of a geometric series, and thus yields

$$p(\nu) = \frac{1}{N} \frac{e^{jN\nu} - 1}{e^{j\nu} - 1} = e^{j(N-1)\nu/2} \frac{\sin(N\nu/2)}{N \sin(\nu/2)}. \tag{5.12}$$

When the elements are driven in phase, $\delta = 0$. In this case, the pattern reaches a maximum at $\theta = 0$, in the direction perpendicular to the array, and the array is called a *broadside* array. If the pattern reaches a maximum at $\theta = \pi/2$, along the direction of the array elements, we have an *endfire* array; the condition for endfire radiation is, therefore, $\delta = -kd$ (e.g., $kd = \pi/2$, $\delta = -\pi/2$).

The array pattern in Eq. (5.12) behaves like a sinc function. The sidelobes of the pattern have a constant width, $\Delta\nu = 2\pi/N$. As we move away from the mainlobe direction ($\nu = 0$), the heights of the sidelobes diminish gradually. Since there are N elements, the pattern has $N-1$ extrema, at which the slope is zero. One extrema is located at the tip of the mainlobe ($\nu = 0$), and the remaining $N-2$ extrema are in the sidelobe region. In Step 1 of the proposed null synthesis method, we place samples at these $N-1$ extrema. Since the slope is zero at these points, they can be obtained by solving

$$\tan(N\nu/2) = \tan(\nu/2). \tag{5.13}$$

However, this is a transcendental equation, and can only be solved numerically. So, instead of doing this, we first compute the pattern at a densely spaced set of points by applying an FFT to the zero-padded weight vector. The extrema of the pattern are then obtained by searching these samples for the local maxima and minima; $N-1$ samples of the pattern are taken at these extrema. The one remaining sample is placed close to the tip of the mainlobe. The aim of doing so is to prevent any overshoot of the pattern around the mainlobe, which might otherwise occur when nulls are placed in the sidelobe region very near the mainlobe. The results obtained are illustrated by design examples in Section 5.4 (Example 5.1).

5.3.2 *Chebyshev Pattern*

Chebyshev polynomials are used to construct linear arrays which are optimum in the sense that they have the narrowest beam width for a prescribed sidelobe level, or the lowest sidelobe level for a prescribed beam width. Consider an N-element array with symmetric excitations,

$$|a_n| = |a_{N-n+1}|, \qquad n = 0, 1, \ldots, \lceil (N-1)/2 \rceil - 1. \tag{5.14}$$

The array pattern can then be expressed as

$$p(\nu) = \begin{cases} \displaystyle\sum_{k=0}^{(N-1)/2} b_k \cos k\nu, & N = \text{odd}, \\ \displaystyle\sum_{k=1}^{N/2} b_k \cos\left(k - \tfrac{1}{2}\right)\nu, & N = \text{even}, \end{cases} \tag{5.15}$$

where $\nu = \delta + kdu$, and the coefficients b_k are related to the excitation coefficients in the following way. For odd N,

$$\begin{aligned} b_0 &= a_{(N-1)/2}, \\ b_k &= 2a_{(N-1)/2-k}, \qquad k = 1, 2, \ldots, (N-1)/2, \end{aligned} \tag{5.16}$$

whereas for even N,

$$b_k = 2a_{N/2-k}, \qquad k = 1, 2, \ldots, N/2. \tag{5.17}$$

The function $\cos kx$ can be expressed as a polynomial of degree k in $\cos x$. Thus, the array pattern in Eq. (5.15) can be written as a polynomial of degree $(N-1)/2$ in $\cos\nu$ when N is odd, or as a polynomial of degree $N-1$ in $\cos(\nu/2)$ for both odd and even N. The former was proposed by Riblet [Riblet, 1947], and the latter by Dolph [Dolph, 1946]. These polynomials are optimized by identifying them with Chebyshev polynomials of the form $T_M(x)$ as defined in Eq. (3.10), where M is the order of the polynomial. With Riblet's formulation, the optimum pattern is given by

$$p(\nu) = \frac{1}{R} T_{(N-1)/2}(a\cos\nu + b), \tag{5.18}$$

where N is odd, the sidelobe level is $1/R$, and a, b are parameters. With Dolph's formulation, the optimum pattern is expressed as

$$p(\nu) = \frac{1}{R} T_{N-1}(c\cos\frac{\nu}{2}), \tag{5.19}$$

where c is a parameter.

The parameters a, b, c in Eq. (5.18) and (5.19) are determined by imposing conditions to obtain the appropriate mapping, depending on the type of array, e.g., broadside or endfire. In general, an N-element Chebyshev array has $N-1$ extrema, one being located at the tip of the mainlobe ($\nu = 0$), and the remaining $N-2$ extrema being in the equiripple sidelobe region. In Step 1 of our null synthesis technique, $N-1$ samples are placed at these extrema, and one sample is placed near the tip of the mainlobe (where $p(\nu) = 1 - 1/R$) so as to control any possible overshoot in the mainlobe when the nulls are placed. Since Chebyshev arrays are constructed from Chebyshev polynomials, we can easily express the extrema in closed form. This simplifies Step 1 considerably. For Riblet's formulation in Eq. (5.18), the $N-2$ pattern extrema in the sidelobe region occur when

$$p(\nu) = \pm\frac{1}{R},$$

or

$$T_{(N-1)/2}(a\cos\nu + b) = \pm 1. \tag{5.20}$$

Simplifying this expression further, we find that, in the part of the visible range, $\delta \leq \nu \leq \delta + kd$, there are $(N-1)/2$ extrema located at

$$\nu_k = \cos^{-1}\left[\frac{1}{a}\left(\cos\frac{2k\pi}{N-1} - b\right)\right], \qquad k = 1, 2, \ldots, (N-1)/2, \tag{5.21}$$

and in the remaining part of the visible range, $\delta - kd < \nu < \delta$, there are $(N-3)/2$ extrema located at

$$\nu_{k+(N-1)/2} = 2\delta - \cos^{-1}\left[\frac{1}{a}\left(\cos\frac{2k\pi}{N-1} - b\right)\right], \qquad k = 1, 2, \ldots, (N-3)/2. \tag{5.22}$$

Similarly, for Dolph's formulation in Eq. (5.19), there are $\lfloor (N-1)/2 \rfloor$ extrema in the region, $\delta \le \nu \le \delta + kd$, located at

$$\nu_k = 2\cos^{-1}\left[\frac{1}{c}\cos\frac{k\pi}{N}\right], \qquad k = 1, 2, \ldots, \lfloor (N-1)/2 \rfloor, \tag{5.23}$$

and $\lceil (N-3)/2 \rceil$ extrema in the region, $\delta - kd < \nu < \delta$, located at

$$\nu_{k+\lfloor (N-1)/2 \rfloor} = 2\delta - 2\cos^{-1}\left[\frac{1}{c}\cos\frac{k\pi}{N}\right], \qquad k = 1, 2, \ldots, \lceil (N-3)/2 \rceil. \tag{5.24}$$

The sample values in Step 1 are obtained by sampling the optimum pattern $p(\nu)$ at the N points described here. The parameters a, b, c, involved in the design of broadside and endfire Chebyshev arrays for Riblet's and Dolph's formulation are determined as follows.

(a) Broadside arrays

The phase difference between adjacent excitation coefficients, δ, is zero for broadside arrays, since the pattern reaches a maximum in the direction $\theta = 0$.

Riblet's formulation gives optimum arrays for an odd number of elements, N. The parameters a and b in Eq. (5.18) for a broadside array [Riblet, 1947] are given by

$$a = \frac{z_0 + 1}{1 + \cos\delta}, \tag{5.25}$$

$$b = \frac{z_0 \cos\delta - 1}{1 + \cos\delta}, \tag{5.26}$$

where

$$\delta = \begin{cases} \pi - kd, & d < \lambda/2, \\ 0, & d \ge \lambda/2. \end{cases} \tag{5.27}$$

Dolph's formulation gives optimum broadside arrays for an even or odd number of elements, but only for $d \ge \lambda/2$. Riblet pointed out that for $d < \lambda/2$, this formulation leads to a nonoptimum design because fewer than the maximum possible number of sidelobes are included in the visible range ($-\pi/2 \le \theta \le \pi/2$ or $\delta - kd \le \nu \le \delta + kd$) [Riblet, 1947]. The parameter c in Eq. (5.19) for a broadside array is given by

$$c = T_N^{-1}(R) = \cosh\left(\frac{1}{N}\cosh^{-1} R\right), \tag{5.28}$$

where $1/R$ is the sidelobe level [Dolph, 1946].

(b) Endfire arrays

For endfire arrays, $d > \lambda/2$ is normally not used, so as to avoid extra mainlobes in the pattern [Pritchard, 1955].

Riblet's formulation gives optimum arrays for an odd number of elements, N. The parameters a, b and δ in Eq. (5.18) for an endfire array are given by

$$a = \frac{-(z_0+3) - 2\cos kd\sqrt{2(z_0+1)}}{2\sin^2 kd}, \tag{5.29}$$

$$b = -1 - a, \tag{5.30}$$

$$\delta = \sin^{-1}\left(\frac{1-z_0}{2a\sin kd}\right), \tag{5.31}$$

$$\delta = \cos^{-1}\left(\frac{z_0+3+2a}{2a\cos kd}\right), \tag{5.32}$$

where

$$z_0 = T_N^{-1}(R) = \cosh\left(\frac{1}{N}\cosh^{-1} R\right), \tag{5.33}$$

and $1/R$ is the sidelobe level [DuHamel, 1953].

Dolph's formulation gives optimum arrays for an even or odd number of elements. The parameters c and δ in Eq. (5.19) for an endfire array are given by

$$c = \frac{\sqrt{z_0^2 + 1 + 2z_0\cos kd}}{\sin kd}, \tag{5.34}$$

$$\delta = 2\tan^{-1}\left\{\cot\frac{kd}{2}\left(\frac{1+z_0}{1-z_0}\right)\right\}, \tag{5.35}$$

where z_0 is as given in Eq. (5.33) [Pritchard, 1955].

5.4 DESIGN EXAMPLES AND COMPARISONS

In this section, we present design examples to demonstrate the proposed NDFT-based method of synthesizing arrays with prescribed nulls, for sinc patterns and Chebyshev patterns. Wherever applicable, the results are compared with those obtained by using the constrained least square technique discussed in Section 5.2. In one of the examples (Example 5.2), we also provide a comparison with the DFT-based null synthesis technique described in Section 5.2. Since the results produced by this method are much worse, it is used in only one example.

To evaluate the performance of these null synthesis methods, we place M nulls over a sector Δu of the pattern, and use two parameters to measure performance—the sidelobe cancellation ratio C, and the gain cost ϵ_g [Steykal, 1982]. The sidelobe cancellation compares the original pattern $p_0(u)$ and the constrained pattern $p_c(u)$ over the desired nulling sector Δu, and is given by

$$C = 20\log_{10}(\max_{u\,\epsilon\,\Delta u} p_0(u)) - 20\log_{10}(\max_{u\,\epsilon\,\Delta u} p_c(u)). \tag{5.36}$$

This measures the depth of the sidelobes in the nulling sector from the original sidelobe level. The gain cost is the decrease in maximum directivity of the pattern in the mainlobe direction u_m, and is given by

$$\epsilon_g = 20\log_{10} p_0(u_m) - 20\log_{10} p_c(u_m). \tag{5.37}$$

For Chebyshev patterns, we also measure the peak sidelobe attenuation in the sidelobe region $u \in u_s$ of the constrained pattern,

$$A_s = -20\log_{10}(\max_{u \in u_s} p_c(u)). \tag{5.38}$$

In the following design examples, a 1024-point FFT was used in Step 4 of the NDFT method.

Example 5.1 Sinc pattern

In this example, we consider the design of two sinc patterns for a broadside array, with $d/\lambda = 0.5$, and the following sets of specifications:

1. An array with $N = 21$ elements, and three nulls at $u = 0.18, 0.22, 0.26$.
2. An array with $N = 41$ elements, and four nulls at $u = 0.22, 0.24, 0.26, 0.28$.

In Table 5.1, we compare the arrays designed by the NDFT method and the constrained least square method. The sidelobe cancellation and gain cost are tabulated. Clearly, the NDFT method gives much better sidelobe cancellation in just two iterations, and zero gain cost.

For the specifications in Set 2, Figs. 5.2(a) and (b) show the constrained sinc pattern designed by the NDFT method, after the first and second iterations. Note that, after the first iteration, the pattern exhibits slight overshoots in the sidelobe regions adjacent to the nulls. These are decreased considerably after

Table 5.1. Performance comparison for null synthesis with sinc patterns.

Set	Method	C (dB)	ϵ_g (dB)
1	NDFT: 1 iteration	30.7037	0
	NDFT: 2 iterations	39.7734	0
	NDFT: 3 iterations	41.0485	0
	NDFT: 4 iterations	41.0626	0
	Least square	36.0331	1.1048
2	NDFT: 1 iteration	33.7851	0
	NDFT: 2 iterations	38.2798	0
	NDFT: 3 iterations	38.8689	0
	NDFT: 4 iterations	38.8728	0
	Least square	33.7619	0.2568

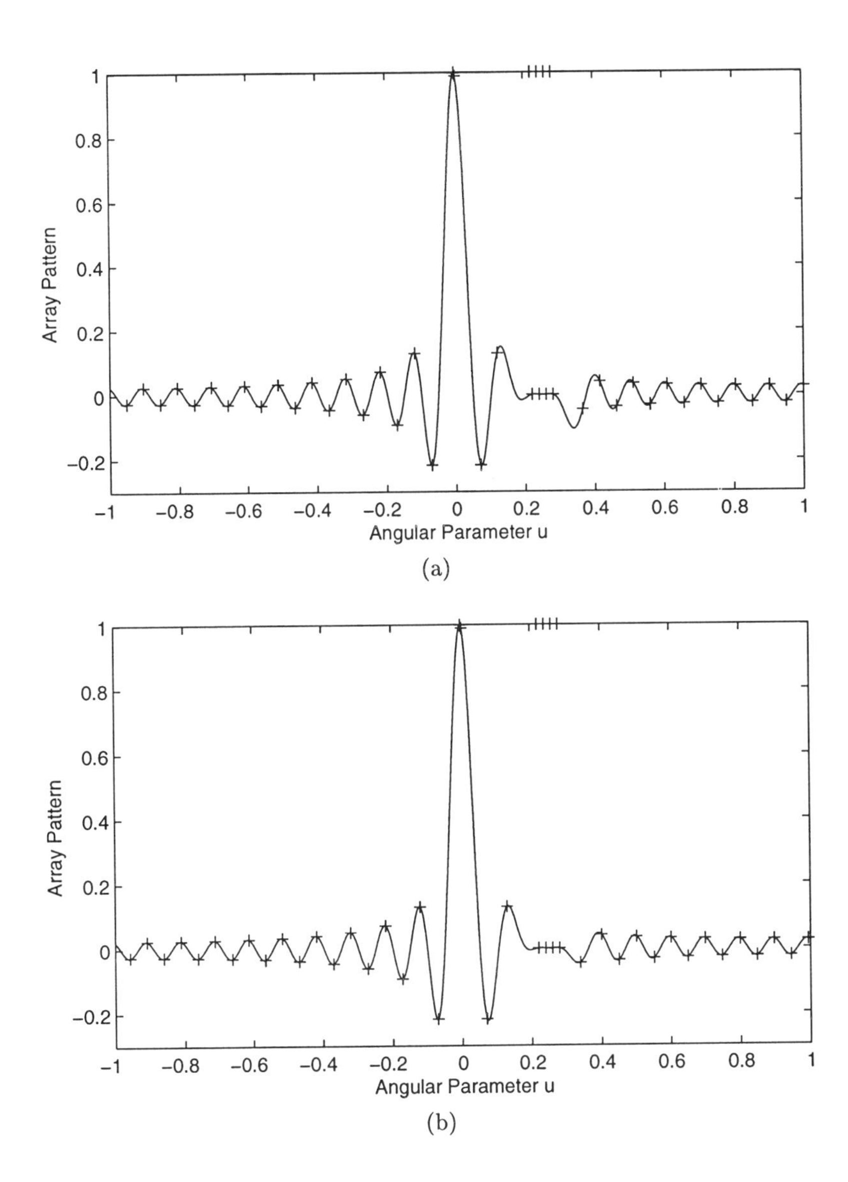

Figure 5.2. Sinc pattern with 41 elements and four nulls, designed by the NDFT method. The sample locations are denoted by "+". (a) Pattern after first iteration. (b) Pattern after second iteration.

the second iteration. The pattern gains obtained by the NDFT method (after two iterations) and the least square method are shown in Figs. 5.3(a) and (b), respectively.

Example 5.2 Chebyshev pattern, broadside array, $d/\lambda = 0.5$

Let us consider the design of two Chebyshev patterns for a broadside array, with $d/\lambda = 0.5$, and the following sets of specifications:

1. An array with $N = 41$ elements, an initial 40-dB Chebyshev pattern, and four nulls at $u = 0.22, 0.24, 0.26, 0.28$.

2. An array with $N = 41$ elements, an initial 40-dB Chebyshev pattern, and eight nulls at $u = 0.22, 0.24, 0.26, 0.28, 0.30, 0.32, 0.34, 0.36$.

Table 5.2 shows the sidelobe cancellation, gain cost and peak sidelobe attenuation in the constrained pattern, obtained by the NDFT method and the constrained least square method. The design times required are shown in the last column. The NDFT method achieves superior sidelobe cancellation in only two or three iterations, with zero gain cost and higher peak sidelobe attenuation.

For the specifications in Set 1, Figs. 5.4(a) and (b) show the Chebyshev pattern and its dB-plot, obtained by the NDFT method after the first iteration. The pattern gains obtained by the NDFT method (after two iterations) and the least square method are shown in Figs. 5.5(a) and (b), respectively. We also compare these to the pattern obtained by using the DFT method. Fig. 5.5(c) shows the resulting pattern after 1000 iterations, using a 1024-point FFT. Table 5.3 tracks the performance of this method with 10, 100, 1000 iterations. Note that even after 1000 iterations, the nulls are not at the desired locations. The sidelobe cancellation obtained is also rather low.

Table 5.2. Performance comparison for null synthesis with Chebyshev patterns for a broadside array ($d/\lambda = 0.5$).

Set	Method	C (dB)	ϵ_g (dB)	$A_s(dB)$	Time (sec)
1	NDFT: 1 iteration	30.2710	0	34.0702	0.3167
	NDFT: 2 iterations	33.8395	0	38.4794	2.2000
	NDFT: 3 iterations	34.1945	0	38.3091	4.0667
	NDFT: 4 iterations	34.1963	0	38.3080	5.8833
	Least square	30.4221	0.0444	34.3527	0.2500
2	NDFT: 1 iteration	43.9043	0	20.1600	0.3167
	NDFT: 2 iterations	51.8774	0	32.2004	2.0167
	NDFT: 3 iterations	55.5224	0	32.2495	3.7000
	NDFT: 4 iterations	55.8117	0	32.0284	5.3833
	Least square	50.5617	0.1674	30.1230	0.4500

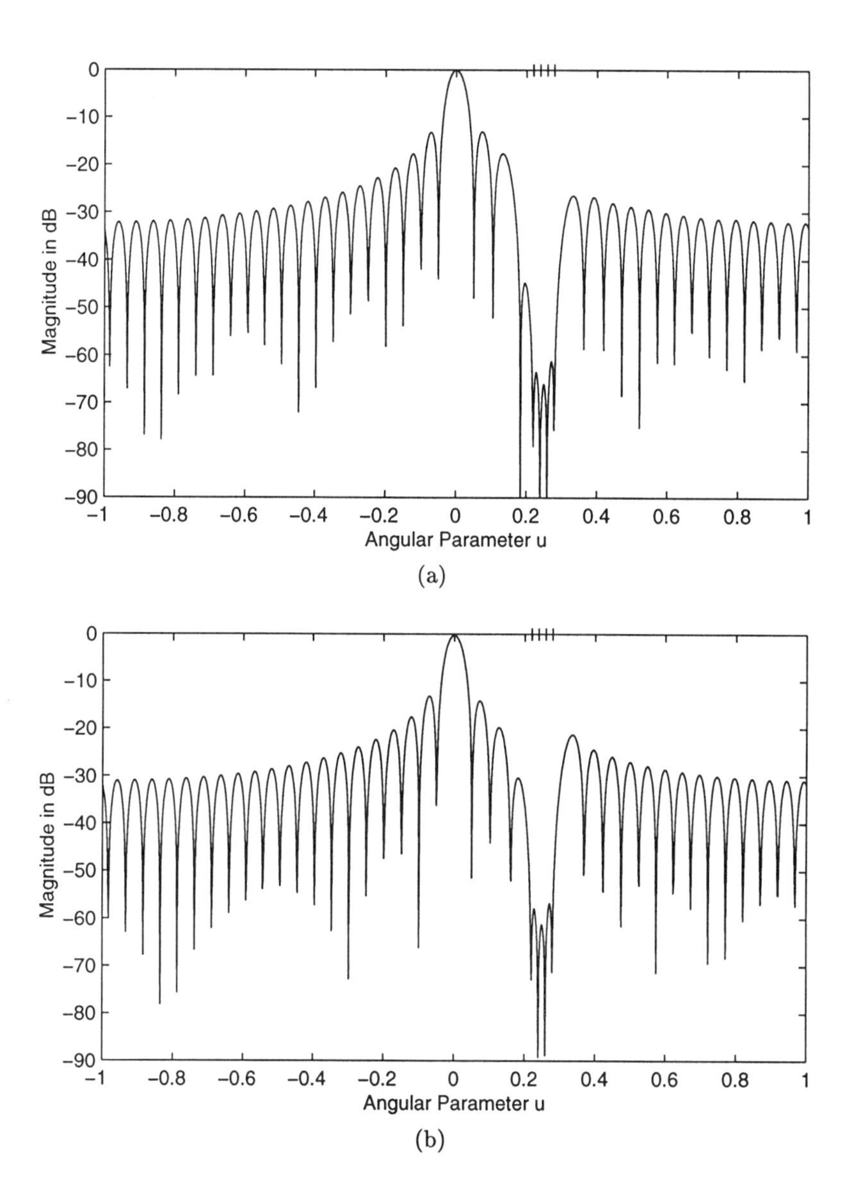

Figure 5.3. Gains of sinc patterns with 41 elements and four nulls, designed by (a) NDFT method (after two iterations), and (b) Least square method.

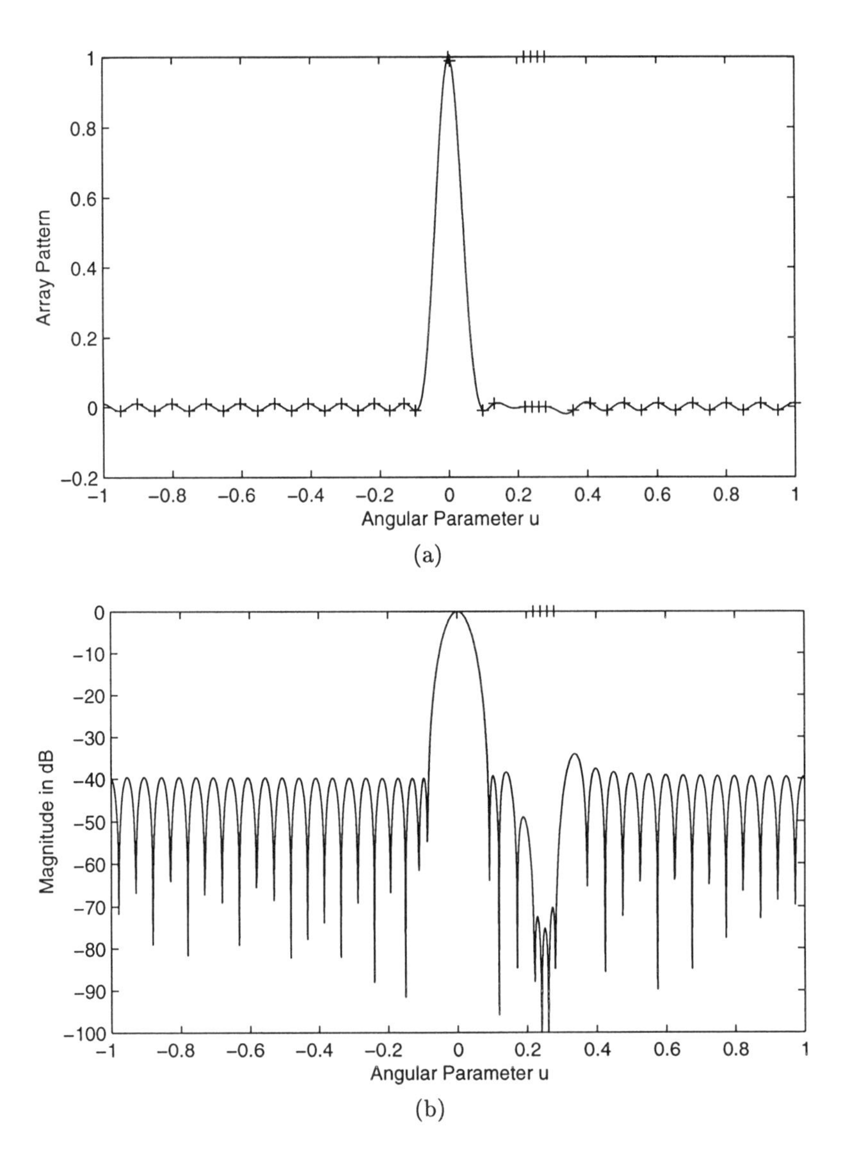

Figure 5.4. Broadside Chebyshev pattern with 41 elements, $d/\lambda = 0.5$, and four nulls, designed by the NDFT method. (a) Pattern after first iteration. The sample locations are denoted by "+". (b) Pattern gain after first iteration.

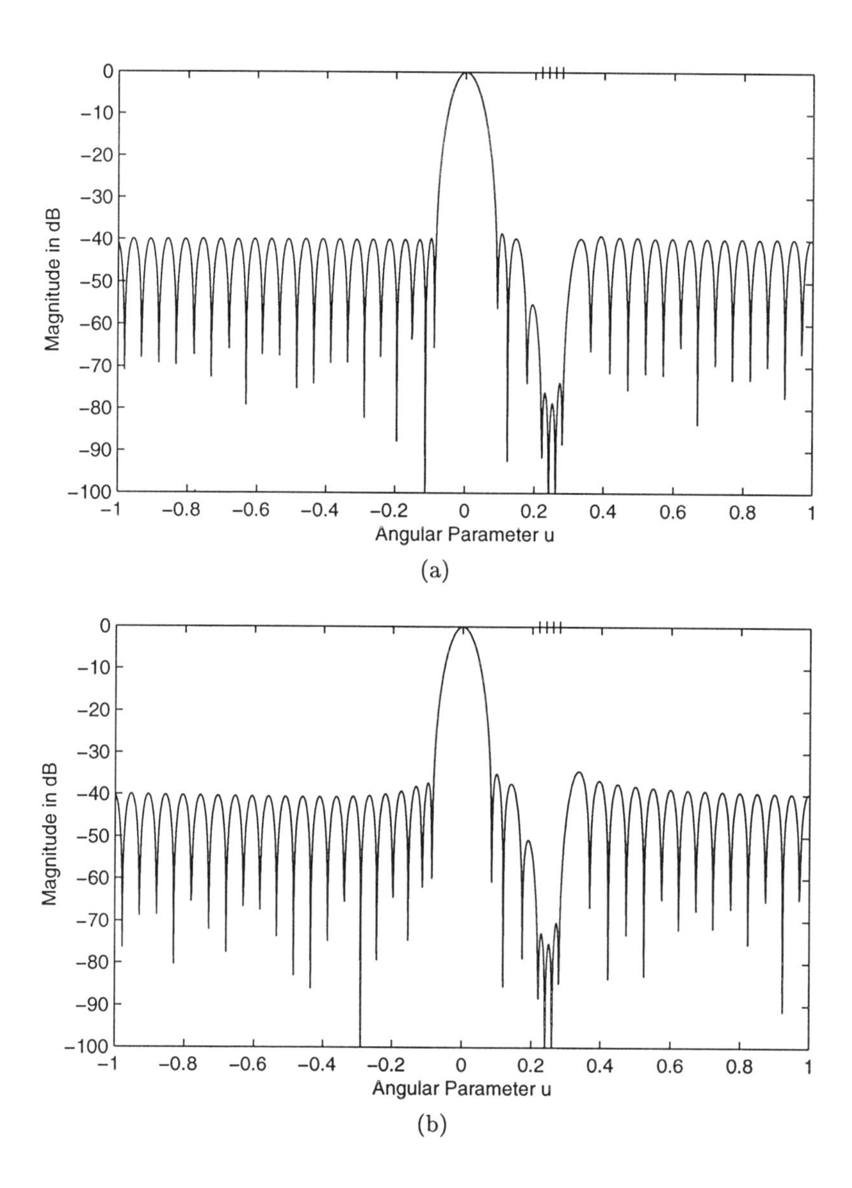

Figure 5.5. Gains of broadside Chebyshev patterns with 41 elements, $d/\lambda = 0.5$, and four nulls, designed by: (a) NDFT method (after two iterations). (b) Least square method.

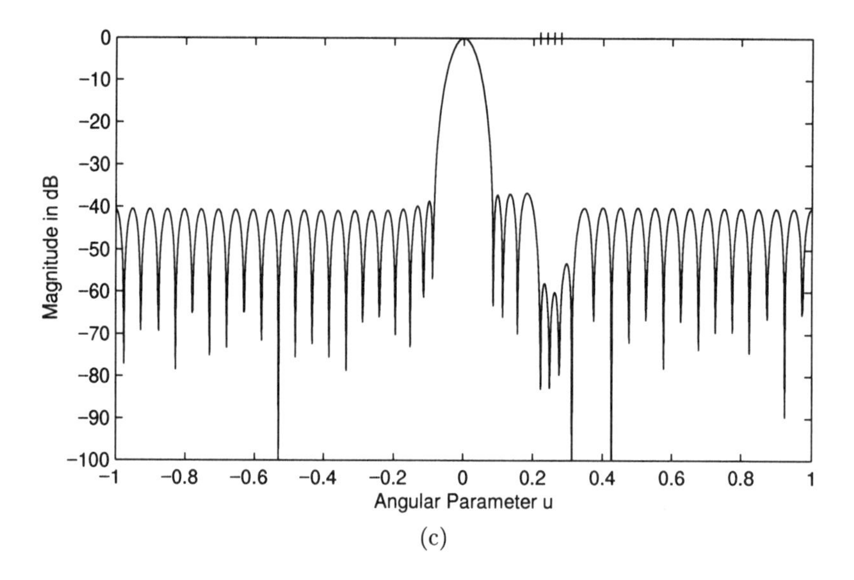

Figure 5.5. (continued) (c) Iterative DFT method after 1000 iterations.

Table 5.3. Performance of the DFT null synthesis method with a Chebyshev pattern for a broadside array ($d/\lambda = 0.5$).

Number of iterations	Nulls obtained	C (dB)	ϵ_g (dB)	A_s (dB)	Time (sec)
10	$u = 0.198, 0.24, 0.281$	3.0624	0.0020	39.3594	3.4167
100	$u = 0.215, 0.24, 0.273$	12.7828	0.0063	37.2010	22.0667
1000	$u = 0.223, 0.248, 0.275, 0.313$	13.2457	0.0107	36.6200	209.6667

Example 5.3 Chebyshev pattern, broadside array, $d/\lambda \neq 0.5$

Now, we consider arrays with interelement spacing other than half-wavelength. Note that the least square solution discussed in Section 5.2 cannot be used in this case. We consider the design of two Chebyshev patterns for a broadside array, using the NDFT method with the following sets of specifications:

1. An array with $N = 47$ elements, $d/\lambda = 0.3$, an initial 40-dB Chebyshev pattern, and three nulls at $u = 0.21, 0.23, 0.25$.

2. An array with $N = 30$ elements, $d/\lambda = 0.6$, an initial 30-dB Chebyshev pattern, and four nulls at $u = 0.20, 0.22, 0.24, 0.26$.

Table 5.4. Performance of the NDFT null synthesis method for a broadside array with Chebyshev patterns, and $d/\lambda \neq 0.5$.

Set	Iterations	C (dB)	ϵ_g (dB)	$A_s(dB)$
1	1	25.5093	0	32.3383
	2	26.1236	0	38.4933
	3	26.3282	0	38.4821
2	1	33.8183	0	24.2554
	2	37.1189	0	28.2585
	3	37.4974	0	28.1144

The values of the sidelobe cancellation, gain cost and peak sidelobe attenuation in the constrained pattern are shown in Table 5.4.

The NDFT method gives good results within just two iterations. The gains of the patterns obtained after two iterations, for the specifications in Sets 1 and 2, are shown in Figs. 5.6(a) and (b).

Example 5.4 Chebyshev pattern, endfire array

Finally, we consider the design of an endfire array with $N = 31$ elements, $d/\lambda = 0.4$, an initial 30-dB Chebyshev pattern, and three nulls at $u = 0.56, 0.58, 0.60$.

Table 5.5 shows the values of the sidelobe cancellation and peak sidelobe attenuation in the constrained pattern obtained by the NDFT method. The gain cost is not relevant since this is an endfire array. The patterns obtained after one and two iterations are shown in Fig. 5.7. By comparing these plots, we note that although the second iteration gives higher sidelobe cancellation, there is also a slight increase in the sidelobe level near $u = -1$, i.e., $\theta = -\pi/2$. This occurs because the Chebyshev pattern has very rapid fluctuations near this

Table 5.5. Performance of the NDFT null synthesis method for an endfire array with a Chebyshev pattern, and $d/\lambda = 0.4$.

Iterations	C (dB)	$A_s(dB)$
1	35.8071	27.1657
2	37.5096	26.1420
3	37.5812	25.8060

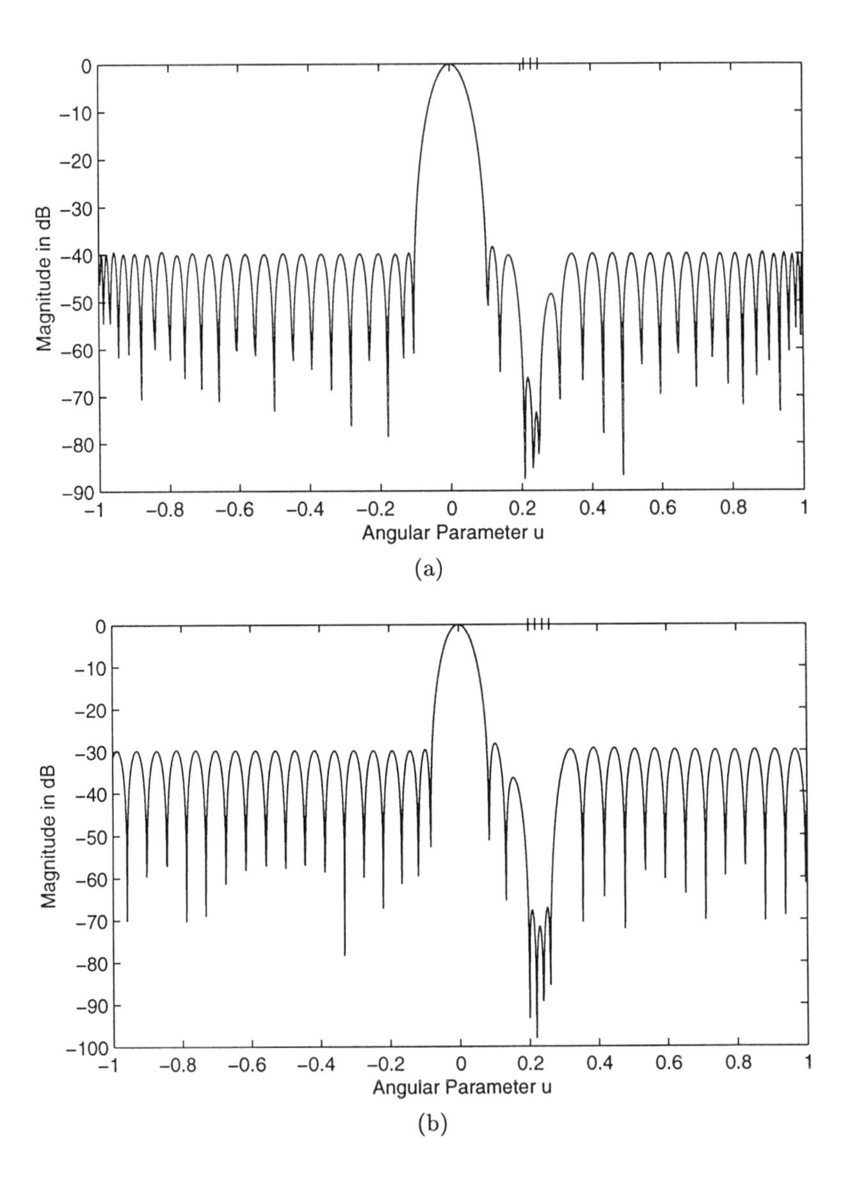

Figure 5.6. Gains of broadside Chebyshev patterns with $d/\lambda \neq 0.5$, and nulls designed by the NDFT method, in Example 5.3. (a) Set 1, (b) Set 2.

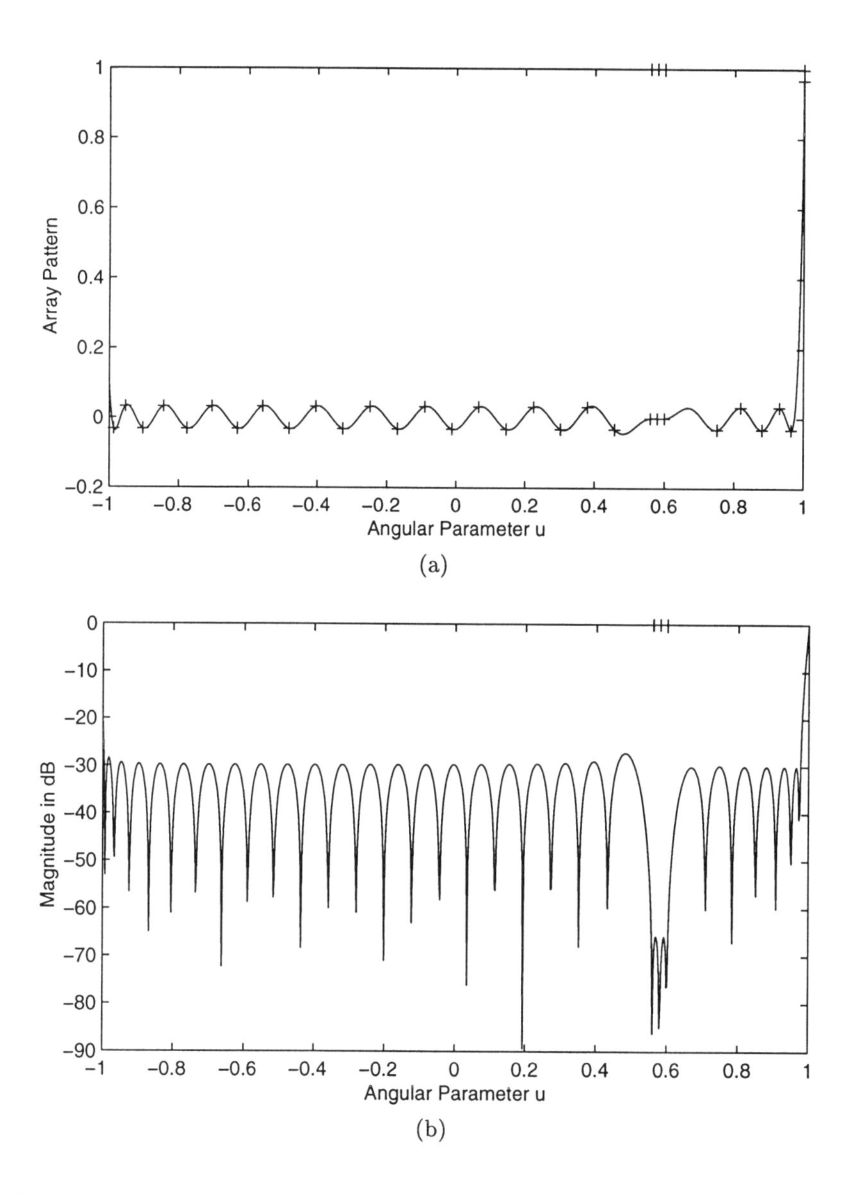

Figure 5.7. Endfire Chebyshev pattern for a 31-element array with $d/\lambda = 0.4$, and three nulls, designed by the NDFT method, in Example 5.4. (a) Pattern after first iteration. The sample locations are denoted by "+". (b) Pattern gain after first iteration.

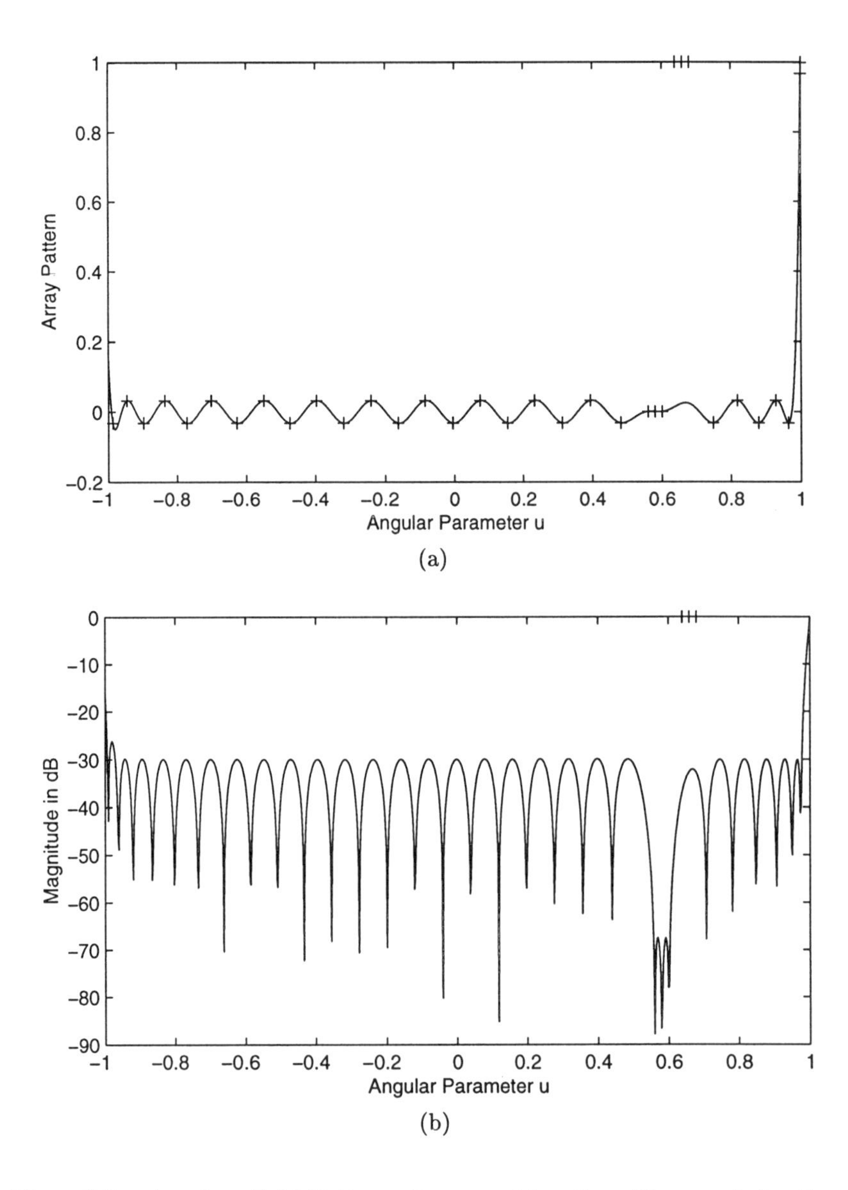

Figure 5.7. (continued) (c) Pattern after second iteration. The sample locations are denoted by "+". (d) Pattern gain after second iteration.

point, and shoots up in value just past it, when $d/\lambda > 0.5$. So it is relatively more difficult to constrain the pattern in this region. Note that this region is also quite far from the main beam direction.

5.5 SUMMARY

In this chapter, we described the application of the NDFT to the problem of designing antenna arrays with nulls located at a set of specified angles. The proposed iterative algorithm for null synthesis was shown to be very effective in achieving high sidelobe cancellation when nulls were placed closely spaced over an angular sector. The effect of the null constraints over the mainbeam and the rest of the sidelobe region was minimal. As illustrated by the design examples, the NDFT-based algorithm produces good results within just two or three iterations. This is significant, since an existing DFT-based algorithm gives much worse results, even after hundreds of iterations. As compared to the constrained least square technique, the NDFT method gives higher attenuation over the nulling sector, with a modest increase in design time. Although we have developed the NDFT method for a linear one-dimensional array with equispaced elements, this approach can be extended to include more general cases such as two-dimensional arrays.

6 DUAL-TONE MULTI-FREQUENCY SIGNAL DECODING

6.1 INTRODUCTION

The dual-tone multi-frequency (DTMF) signaling system is being used in push-button telephone sets worldwide as an improvement over dial-pulse signaling in rotary telephones because of the high dialing speed which can be obtained. Besides telephone call signaling, DTMF is being increasingly used for interactive control applications such as voice mail, electronic mail, and telephone banking. Typically, in such applications, the user selects options from a menu by sending DTMF signals from the keypad of a push-button telephone.

A DTMF signal consists of the sum of two tones, whose frequencies are taken from two mutually exclusive groups of pre-assigned frequencies. To decode a DTMF signal, we must extract these two tones, and thus, find the digit that was dialed. In this chapter, we apply the NDFT for decoding DTMF signals. The two tones are extracted by employing the NDFT to evaluate the spectrum of the DTMF signal at the specified frequencies. Specifically, we use the *subband NDFT*, which was introduced in Chapter 2 (Section 2.4), to develop a new algorithm for DTMF tone detection [Bagchi and Mitra, 1995a; Bagchi and Mitra, 1997]. By utilizing the information that DTMF frequencies are assigned to the *low-frequency* part of the telephone bandwidth, we can discard calculations in the high-frequency band of a two-band SB-NDFT, and consequently, achieve significant savings in computation without sacrificing performance. We present a performance comparison between DTMF detection algorithms based on the DFT, SB-DFT, NDFT, and SB-NDFT. Simulations show that the SB-NDFT requires the *lowest number of computations* to achieve a specified level of performance. Since multi-channel DTMF decoding is performed by time-multiplexing the channels on a DSP, a lower decoding time implies that a larger number of DTMF channels can be decoded simultaneously. This is desirable in multi-channel environments such as T-1 facilities (24 channels).

This chapter is organized as follows. Section 6.2 covers necessary background information—DTMF standards, digital implementation of DTMF generation and decoding on DSPs, and a review of existing decoding algorithms based on the DFT and NDFT. In Section 6.3, we propose a new algorithm for DTMF decoding using the SB-NDFT. The results of simulations are presented in Section 6.4, including a performance comparison between the algorithms.

6.2 BACKGROUND

6.2.1 DTMF Standards

The standards for DTMF signaling with push-button telephone sets appear as Recommendations Q.23 and Q.24 in the CCITT Red Book [International Telecommunication Union, 1984, Section 4.3]. This is an in-band system because the signaling frequencies are located in the frequency band used for speech.

In Recommendation Q.23, frequencies are allocated to the various digits and symbols of a push-button keypad, as shown in Fig. 6.1. Note that the four keys

Row ↓ \ Column →	1209 Hz	1336 Hz	1477 Hz	1633 Hz
697 Hz	1	2	3	A
770 Hz	4	5	6	B
852 Hz	7	8	9	C
941 Hz	*	0	#	D

Figure 6.1. Allocation of frequencies for a push-button keypad.

Table 6.1. Values of DTMF receiving parameters adopted by AT&T for the TouchTone standard.

Parameters		Values
Signal frequencies	Low group	697, 770, 852, 941 Hz
	High group	1209, 1336, 1477, 1633 Hz
Frequency tolerance $\lvert\Delta f\rvert$	Operation	$\leq 1.5\%$
	Non-operation	$\geq 3.5\%$
Power levels per frequency	Operation	0 to −25 dBm
	Non-operation	Max. −55 dBm
Power level difference between frequencies		+4 dB to −8 dB
Signal duration	Operation	Min. 40 ms
	Non-operation	Max. 23 ms

in the last column do not appear on standard keypads, and are reserved for future use. Whenever a button is pressed, a DTMF signal containing *two frequencies* is generated. These two frequencies belong to two mutually exclusive frequency groups of four frequencies each, as shown in Fig. 6.1. One frequency corresponds to the *row*, and the other corresponds to the *column* of the digit dialed. This code is known as the "2(1/4) code".

Recommendation Q.24 specifies operational values for several technical parameters, applicable for the reception of DTMF signals in local exchanges. Since factors such as transmission loss vary among national networks, various administrations have slightly different standards. Table 6.1 shows the relevant parameters in the AT&T standard, *TouchTone*, since they are important to ensure proper decoding of DTMF signals under extreme receiving conditions. As specified in Recommendation Q.24, the DTMF receiver must provide a check for the simultaneous presence of one and only one row frequency and one and only one column frequency. DTMF receivers are required to detect operational frequencies with a *tolerance* of $\pm 1.5\%$. The tolerances for the *power levels* in Table 6.1 indicate that there must be a minimum difference of $-25 - (-55) = 30$ dBm between the levels of the operational tones and the nonoperational tones. The power difference between the two operational tones is known as *twist*. Normal twist occurs when the received level of the column tone is lower than the row tone, due to attenuation of high frequencies by the telephone network. Similarly, reverse twist occurs when the column tone level is higher than the row tone level. The maximum allowable twist values are 8 dB for normal twist and 4 dB for reverse twist.

6.2.2 Digital Implementation

Although several chips with analog circuitry are available for the generation and decoding of DTMF signals in a single channel, these functions can also be implemented digitally on DSPs. Such a *digital system* surpasses analog equivalents in performance, since it provides *better* precision, stability, versatility, reprogrammability, and the scope for multi-channel operation leading to lower chip count. Examples of digital implementation use DSPs such as the TMS32010 [Mock, 1985], WE DSP-32 [Gay et al., 1989], and ADSP-2100 [Marr, 1990]. Codec (coder/decoder) chips or linear analog-to-digital (A/D) converters and digital-to-analog (D/A) converters provide the required analog interface.

The digital *generation* of a DTMF signal is simple, since it involves adding digital samples of two sinusoids, which are generated by using look-up tables or by computing a polynomial expansion.

The *decoding* of DTMF signals, which is the focus of this chapter, is a more intricate problem. In analog circuits, tone detection is often performed by detecting and counting zero-crossings of the input signal. In a digital implementation, tone detection can be performed easily by transforming the input samples to the frequency domain by means of the Fourier transform. This *transformation* can be accomplished in several ways, as discussed later in this chapter. However, each method results in a set of frequency-domain samples which measure the energy present at the eight DTMF frequencies. The next task is to determine whether the tones detected constitute a valid DTMF digit. This is done by performing several *digit validation* tests, such as:

1. Magnitude test: The largest row tone and the largest column tone must each greater than a certain threshold.

2. No-other-tones test: This test checks the levels of the row and column tones other than the largest. The result of dividing the magnitude of each such row (or column) tone by the magnitude of the largest row (or column) tone must be lower than a specified threshold.

3. Twist test: The magnitudes of the largest row and column tones are compared to each other by dividing the magnitude of the row tone by the magnitude of the column tone. For normal twist, this ratio should be lower than a certain threshold; for reverse twist, it should be higher than a specified threshold.

4. Tone-to-total energy test: The total energy present is calculated and compared to the sum of the energies at the largest row and column tones. This prevents a digit from being simulated by speech. For a DTMF signal, these two values are equivalent. However, for speech, the total energy is much greater.

5. Harmonic test: This also serves as a check against digit simulation. The energies at the second harmonics of the largest row and column tones are

computed. They are negligible for a DTMF signal, but usually high for speech.

If all these tests pass, the digit corresponding to the dialed tones is decoded. Usually, a valid digit is declared if the digits decoded from two successive frames of the signal are the same.

The decoding operation is, therefore, performed in two steps—transformation and digit validation. In the remaining part of this chapter, we shall focuss on the transformation used for converting the signal to the frequency domain. Let N be the number of samples required to properly detect all DTMF digits under the receiving conditions specified by the AT&T standard in Table 6.1. Since the minimum duration of a DTMF signal is 40 ms, and the sampling rate is 8 kHz, the number of samples available for decoding each DTMF digit is $0.04 \times 8000 = 320$. Thus, we have

$$N \leq 320. \tag{6.1}$$

The value of N actually required depends on the transformation method used. Once N has been found, the thresholds required for digit validation can be found by determining the worst-case signal levels under extreme receiving conditions. We compare the performances of four transformation methods in this chapter. Two existing DTMF decoding algorithms based on the DFT and NDFT, are outlined in Section 6.2.3. We propose a new algorithm based on the SB-NDFT in Section 6.3.

6.2.3 Existing DTMF Decoding Algorithms

The DFT has been used to measure the energy of the received DTMF signal at the eight tone frequencies. Such DFT-based DTMF decoder implementations include those for the TMS32010 [Mock, 1985] and the ADSP-2100 [Marr, 1990]. They compute the eight DFT samples $X[k]$ which are closest in frequency to the eight DTMF tones. The Goertzel algorithm requires one second-order recursive filter for each frequency, and is used for efficient implementation [Mitra, 1998]. The computation is slightly modified because we need only the squared magnitudes $|X[k]|^2$. This algorithm is a special case of the modified Goertzel algorithm discussed in Chapter 2 (Section 2.3.3, page 33), with $\omega_k = 2\pi k/N$. The problem of using the DFT is that the actual DTMF frequencies do not coincide exactly with the equispaced DFT samples. An *error* is introduced by this *mismatch* in sampling.

In an improved algorithm, this problem has been overcome by using samples at the actual DTMF frequencies. This approach is equivalent to evaluating the NDFT samples at these frequencies, and was used in an implementation for the WE DSP32 family [Gay et al., 1989]. The square magnitudes of the NDFT samples are computed by using the modified Goertzel algorithm discussed in Chapter 2. Since there is *no frequency mismatch*, the NDFT-based algorithm performs better than the DFT, as illustrated later in Section 6.4.

As we showed in Chapter 2, each of these algorithms (based on the DFT and NDFT) require $(N+4)$ real multiplications and $(2N+2)$ real additions for each frequency, leading to a total of $(8N + 32)$ real multiplications and $(16N + 16)$ real additions for the eight DTMF frequencies.

6.3 PROPOSED DTMF DECODING ALGORITHM USING THE SUBBAND NDFT

We propose a new algorithm for DTMF decoding based on the subband NDFT (SB-NDFT), which was introduced in Chapter 2 [Bagchi and Mitra, 1995a; Bagchi and Mitra, 1997]. Since the sampling rate used for telephony is $f_s =$ 8 kHz, all the DTMF frequencies (as shown in Table 6.1) are located in the low-frequency band $0 \leq f \leq f_s/4$. Utilizing this information, we can obtain a *good approximation* to the NDFT samples by using a two-band SB-NDFT and discarding calculations for the high-frequency band. This *reduces the computation* by nearly half.

Consider a sequence $x[n]$ with an even number of samples N. As discussed in Chapter 2 (Section 2.4), we can decompose $x[n]$ into two subsequences, $g_L[n]$ and $g_H[n]$, of length $N/2$ each:

$$g_L[n] = \frac{1}{2}\{x[2n] + x[2n+1]\}, \tag{6.2}$$

$$g_H[n] = \frac{1}{2}\{x[2n] - x[2n+1]\} \tag{6.3}$$

$$n = 0, 1, \ldots, N/2 - 1.$$

We can express the NDFT of $x[n]$ as

$$X(z_k) = (1 + z_k^{-1})\, G_L(z_k^2) + (1 - z_k^{-1})\, G_H(z_k^2), \tag{6.4}$$

where $G_L(z_k^2)$ and $G_H(z_k^2)$ are the NDFTs of the subsequences, $g_L[n]$ and $g_H[n]$, evaluated at $z = z_k^2$. Since the eight DTMF tones are located in the low-frequency band, we can obtain a reasonable approximation to the NDFT in Eq. (6.4) by dropping the highpass term $G_H(z_k^2)$ to get

$$\hat{X}(z_k) = (1 + z_k^{-1})\, G_L(z_k^2), \qquad k = 0, 1, \ldots, 7, \tag{6.5}$$

where $z_k = e^{j\omega_k}$, $\omega_k = 2\pi f_k / f_s$, and $f_0, f_1, \ldots, f_7$ are the eight DTMF frequencies. Note that the amount of computation required in Eq. (6.5) is nearly *half* of that in Eq. (6.4). The dropping of the higher subband results in linear distortion of the NDFT samples due to a non-constant frequency response in the low-frequency band. As derived in Eq. (2.110) of Chapter 2, we can compensate for this distortion by scaling

$$X(e^{j\omega_k}) = 2b_k\, \hat{X}(e^{j\omega_k}), \tag{6.6}$$

where the parameter b_k is defined as

$$b_k = \frac{1}{1 + \cos\omega_k}. \tag{6.7}$$

For DTMF decoding, we need only the *squared magnitudes* of the approximate NDFT samples. Squaring both sides of Eq. (6.6), we have

$$|X(e^{j\omega_k})|^2 = 4{b_k}^2 \, |\hat{X}(e^{j\omega_k})|^2. \tag{6.8}$$

Using Eq. (6.5), we obtain

$$\begin{aligned} |\hat{X}(e^{j\omega_k})|^2 &= |(1+e^{-j\omega_k})|^2 \, |G_L(e^{j2\omega_k})|^2 \\ &= \frac{2}{b_k} \, |G_L(e^{j2\omega_k})|^2. \end{aligned} \tag{6.9}$$

From Eqs. (6.8) and (6.9), we arrive at

$$|X(e^{j\omega_k})|^2 = 8b_k \, |G_L(e^{j2\omega_k})|^2. \tag{6.10}$$

The value of $|G_L(e^{j2\omega_k})|^2$ is computed by using the modified Goertzel algorithm. A brief outline of this procedure is as follows. Let us express $G_L(z_k^2)$ as

$$G_L(z_k^2) = \sum_{r=0}^{N/2-1} g_L[r] z_k^{-2r} = z_k^{-N} \, y_k[n]|_{n=N/2}, \tag{6.11}$$

where $y_k[n]$ is the result of the discrete convolution

$$y_k[n] = g_L[n] * z_k^{2n} u[n] = \sum_{r=0}^{N/2-1} g_L[r] z_k^{2(n-r)} u[n-r], \tag{6.12}$$

with $u[n]$ denoting the unit step sequence. This is equivalent to a second-order recursive filter with a system function

$$H_k(z) = \frac{1}{1 - z_k^2 z^{-1}} = \frac{1 - e^{-j2\omega_k} z^{-1}}{1 - a_k \, z^{-1} + z^{-2}}. \tag{6.13}$$

where $z_k = e^{j\omega_k}$, and the parameter a_k is defined as

$$a_k = 2\cos 2\omega_k. \tag{6.14}$$

This system is depicted in Fig. 6.2. We can compute $y_k[N/2]$ by solving for the intermediate signals $q_k[n]$ given by the difference equations

$$q_k[n] = a_k \, q_k[n-1] - q_k[n-2] + g_L[n], \qquad n = 0, 1, \ldots, \infty, \tag{6.15}$$

with the initial conditions

$$q_k[-1] = q_k[-2] = 0.$$

Since $g_L[n] = 0$ for $n < 0$ and $n \geq N/2$, it follows that

$$y_k[N/2] = q_k[N/2] - e^{-j2\omega_k} \, q_k[N/2-1]. \tag{6.16}$$

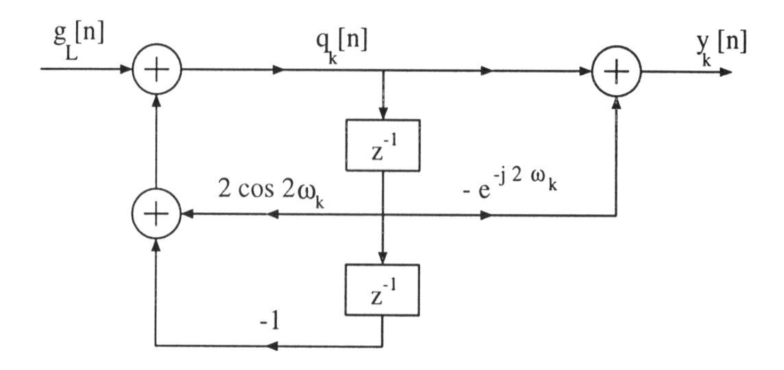

Figure 6.2. Goertzel algorithm as a second-order recursive computation.

From Eq. (6.11), we obtain

$$|G_L(e^{j2\omega_k})|^2 = |y_k[N/2]|^2. \tag{6.17}$$

From Eq. (6.16),

$$|y_k[N/2]|^2 = q_k^2[N/2] + q_k^2[N/2-1] + a_k\, q_k[N/2]\, q_k[N/2-1]. \tag{6.18}$$

Combining Eqs. (6.17) and (6.18), the required squared magnitude is given by

$$|G_L(e^{j2\omega_k})|^2 = q_k^2[N/2] + q_k^2[N/2-1] + a_k\, q_k[N/2]\, q_k[N/2-1]. \tag{6.19}$$

Fig. 6.3 shows the operations involved in the proposed algorithm. Note that the factor of half in Eq. (6.2), and the factor of eight in Eq. (6.10) can be neglected in the calculation because only the relative squared magnitudes of $|X(e^{j\omega_k})|^2$ for the eight DTMF tones are of importance. In this algorithm, we need only two coefficients, a_k and b_k, for each DTMF frequency. As shown in Fig. 6.3, a_k is a multiplier used in the Goertzel algorithm, and b_k is a scaling factor for the output.

To find the computational complexity of the proposed algorithm, let us summarize the steps involved:

1. Obtain the subsequence $g_L[n]$ from the input samples $x[n]$. This requires $N/2$ real additions.

2. For each DTMF tone, compute the squared magnitude of the NDFT of $g_L[n]$, evaluated at twice the tone frequency, using the modified Goertzel algorithm. This requires $(N/2+4)$ real multiplications and $(N+2)$ real additions.

3. For each DTMF tone, multiply the squared magnitude by the scaling factor b_k. This requires one real multiplication.

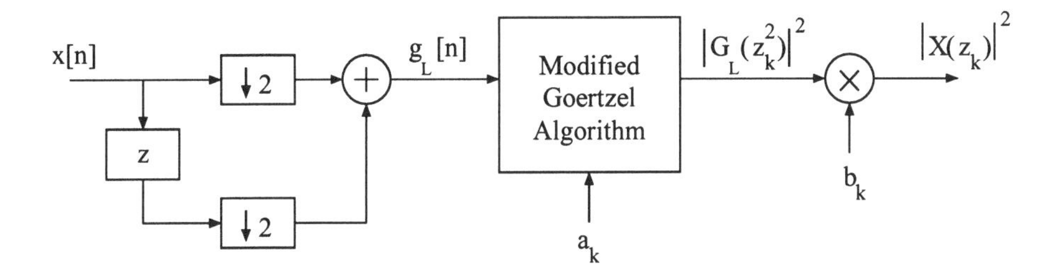

Figure 6.3. DTMF decoding algorithm using the SB-NDFT.

Therefore, the total computation required for the eight DTMF tones is $(4N+40)$ real multiplications and $(8.5N + 16)$ real additions. Since existing algorithms based on the DFT [Mock, 1985] and NDFT [Gay et al., 1989] require $(8N + 32)$ real multiplications and $(16N + 16)$ real additions for the eight tones, the proposed algorithm based on the SB-NDFT has a much lower computational complexity (nearly *half*).

Note that we can also have a DTMF decoding algorithm based on the SB-DFT. This is similar to the SB-NDFT algorithm described here, except that $\omega_k = 2\pi k/N$. Eight integer values are chosen for k such that the corresponding frequencies kf_s/N are closest to the DTMF frequencies. Although the computational complexity of this algorithm is the same as for the SB-NDFT, the sampling error caused by uniform sampling leads to a worse performance. This is illustrated in Section 6.4.

6.4 RESULTS AND COMPARISONS

In this section, our aim is to find the minimum number of input samples, N, required to properly detect DTMF digits under the receiving conditions specified by the TouchTone standard. By simulating these conditions, we compare the performances of four DTMF decoding algorithms based on the NDFT, SB-NDFT, DFT, and SB-DFT, respectively.

We start by defining a figure of merit, which is used as a performance measure. Then, we discuss factors affecting the performance of DTMF detection algorithms. An understanding of these factors is essential for interpreting the performances of the detection algorithms. Finally, we present the results of simulations, which have been used to find the minimum number of samples required by each algorithm to attain a desired figure of merit.

6.4.1 Figure of Merit

To compare the performance of the proposed algorithm with those of other approaches, we define a new measure. Assuming that the dialed DTMF digit was decoded successfully, the figure of merit corresponds to the minimum among the differences between the power levels (dB) of the highest and second-highest row and column tones:

$$F = \min(D_r, D_c), \tag{6.20}$$

where

$$\begin{aligned} D_r &= 10\log_{10} X_r - 10\log_{10} X_{rs}, \\ D_c &= 10\log_{10} X_c - 10\log_{10} X_{cs}. \end{aligned} \tag{6.21}$$

Here, X_r and X_c are the maximum values among the energies detected at the row and column tones, respectively; X_{rs} and X_{cs} are the second-highest values among the energies detected at the row and column tones, respectively. Clearly, a positive value for the figure of merit ($F > 0$) is required for successful decoding. A higher figure of merit is desirable so that the operational tones are detected at levels reliably higher than the non-operational tones. Typically, the number of input samples is chosen so that the worst-case figure of merit is about 6 dB.

6.4.2 *Factors Affecting Performance of Detection Algorithms*

Now, we briefly discuss the factors affecting the performance of DTMF detection algorithms. The DTMF signal consists of the sum of two sinusoids. The input to the detection algorithm is a windowed segment of the sampled signal, given by

$$v[n] = x[n]\, w[n], \tag{6.22}$$

where

$$x[n] = A_r \cos\omega_r n + A_c \cos\omega_c n, \tag{6.23}$$

and

$$w[n] = \begin{cases} 1, & 0 \le n \le N-1, \\ 0, & \text{otherwise.} \end{cases} \tag{6.24}$$

Here, A_r and A_c are the amplitudes, and ω_r and ω_c are the frequencies of the row and column sinusoids, respectively. The sequence $w[n]$ in Eq. (6.24) is a rectangular window [Mitra, 1998], with a discrete-time Fourier transform

$$W(e^{j\omega}) = \sum_{n=0}^{N-1} e^{-j\omega n} = e^{-j\omega(N-1)/2}\frac{\sin(\omega N/2)}{\sin(\omega/2)}. \tag{6.25}$$

The discrete-time Fourier transform of a sinusoid $A\cos\omega_0 n$ is a pair of impulses at $+\omega_0$ and $-\omega_0$. The effect of windowing can be studied by considering the Fourier transform $V(e^{j\omega})$ of the windowed sequence $v[n]$. By expanding Eq. (6.23) in terms of complex exponentials, and utilizing the frequency shifting property of the Fourier transform, we obtain

$$\begin{aligned} V(e^{j\omega}) &= \frac{A_r}{2}W(e^{j(\omega-\omega_r)}) + \frac{A_r}{2}W(e^{j(\omega+\omega_r)}) \\ &\quad + \frac{A_c}{2}W(e^{j(\omega-\omega_c)}) + \frac{A_c}{2}W(e^{j(\omega+\omega_c)}), \end{aligned} \tag{6.26}$$

which consists of the Fourier transform of the window replicated at the frequencies $\pm\omega_r$ and $\pm\omega_c$ and scaled by the amplitudes A_r and A_c, of the corresponding sinusoids, respectively. The figure of merit obtained depends on the receiving conditions, and on the number of samples N.

The relative values of the amplitudes of the row and column tones in the received DTMF signal depend on the existing twist, which is the power difference between the two operational tones. To consider the effect of *twist*, suppose the digit "1" is dialed, and the frequencies $f_r = 697$ Hz and $f_c = 1209$ Hz are generated. Figs. 6.4(a), (b), and (c) show the magnitude of the Fourier transform $V(e^{j\omega})$, with $N = 100$, under three twist conditions—no twist ($A_r = A_c$), maximum reverse twist (A_c higher than A_r by 4 dB) and maximum normal twist (A_c lower than A_r by 8 dB), respectively. From these plots, it is clear that twist affects performance markedly. Usually, the lowest figure of merit is obtained for maximum normal twist. This occurs because the height of the mainlobe at the column tone is lowered substantially, bringing it closer to the high sidelobes of the row tone. Consequently, high values are detected at the nonoperational column tones, and the figure of merit is decreased.

As we increase the *number of input samples* N, the width of the mainlobe ($\Delta\omega = 4\pi/N$) at each operational tone decreases. This is illustrated in Fig. 6.5 for $N = 320$, which is the maximum number of samples available for this application, as was shown in Eq. (6.1). On comparing these plots with those for $N = 100$ in Fig. 6.4, we see that for $N = 320$, the sidelobes have been pulled towards the mainlobe—this is desirable since lower values are now detected at the nonoperational frequencies, and the figure of merit is increased.

However, there is an *upper bound* on the number of input samples that can be actually used without decreasing the worst-case figure of merit. This restriction arises due to the *frequency tolerance* specified in the standard. The DTMF receiver is required to detect operational tones within bands that are ± 1.5 % about the DTMF frequencies. Now, if we are are sampling the spectrum at the center of a band and the corresponding tone moves away towards the edge of that band, we cannot detect the presence of the tone if the mainlobe is too narrow, i.e., when N is too large. This situation is illustrated by an example in Fig. 6.6. In this case, $N = 320$ and the digit "*" has been dialed. The column tone occurs at the lower extreme of the eighth band. Since the sample taken at the center of this band falls outside the mainlobe, we fail to detect the operational column tone. Therefore, we infer that increasing the number of samples causes the detector performance to improve initially, but to deteriorate after a certain point. This is also illustrated in Fig. 6.7, which shows that as the number of samples is increased, the figure of merit increases initially but decreases after a certain point.

Finally, let us consider the effect of the presence of *nonoperational tones*, whose levels are at least 30 dB below the lower of the two main tones. Fig. 6.7 shows an example when the digit "1" is dialed. For different values of N, the number of input samples, the figure of merit F is obtained by computing samples of the windowed Fourier transform of the DTMF signal at the eight DTMF

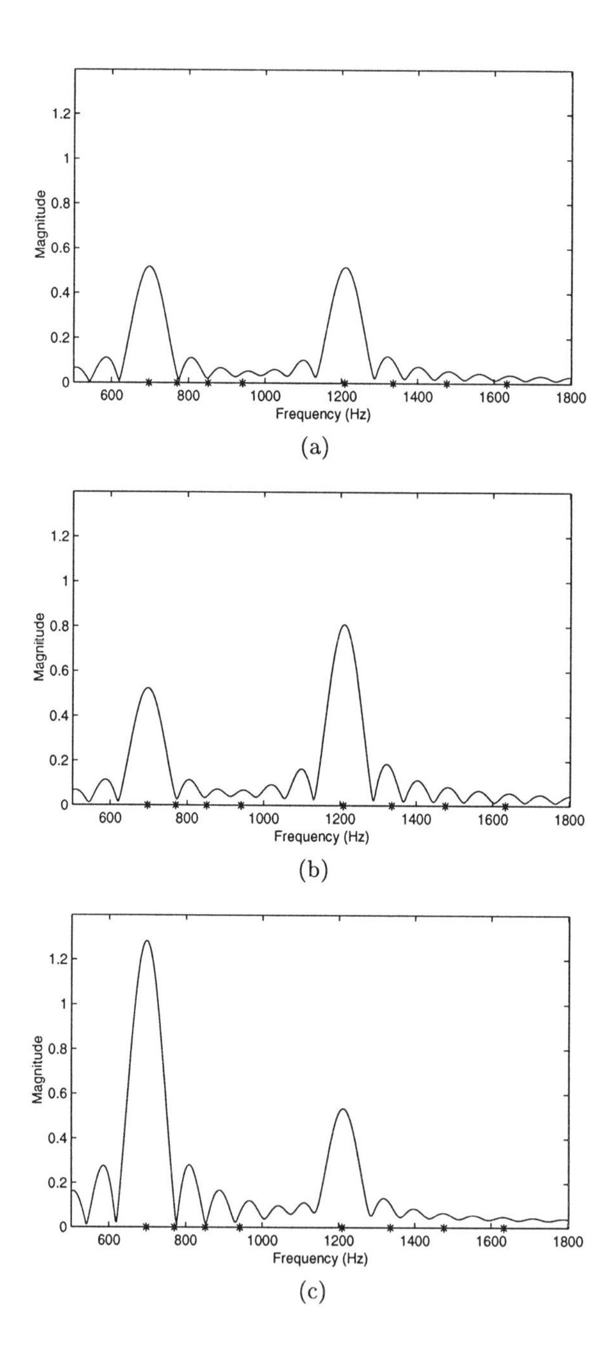

Figure 6.4. Magnitude of the Fourier transform of a DTMF signal with 100 samples, generated when the digit "1" is dialed. (The eight DTMF frequencies are denoted by "$*$" on the horizontal axis.) (a) No twist. (b) Reverse twist of 4 dB. (c) Normal twist of 8 dB.

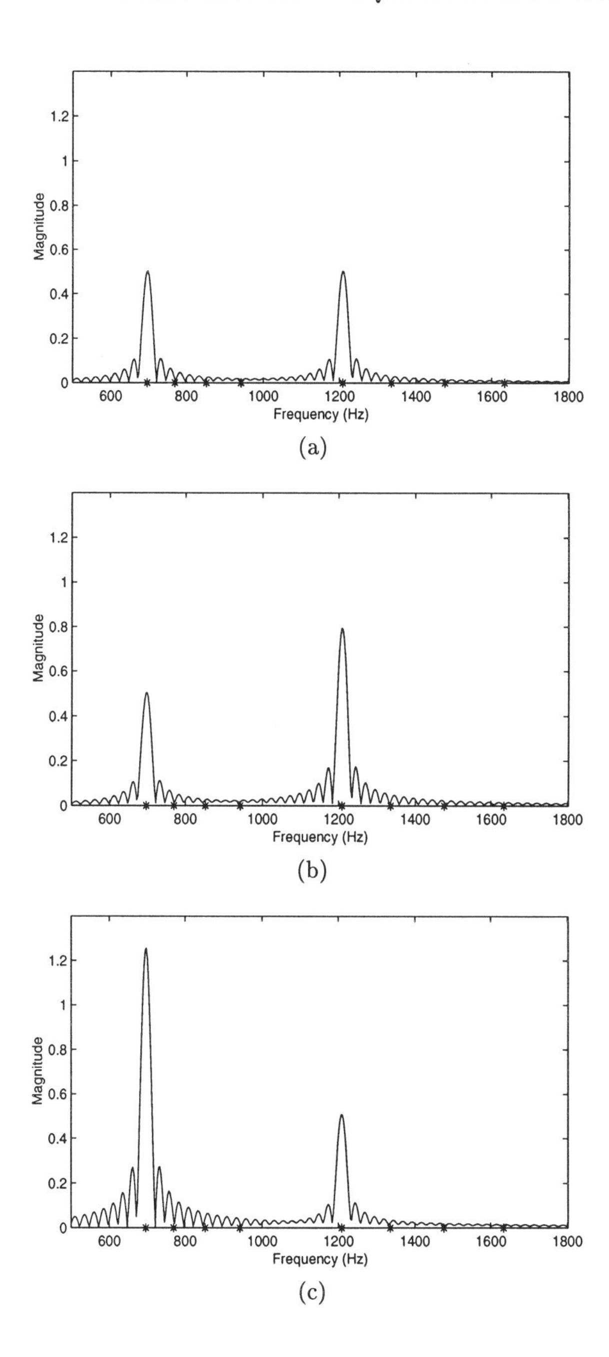

Figure 6.5. Magnitude of the Fourier transform of a DTMF signal with 320 samples, generated when the digit "1" is dialed. (The eight DTMF frequencies are denoted by "*" on the horizontal axis.) (a) No twist. (b) Reverse twist of 4 dB. (c) Normal twist of 8 dB.

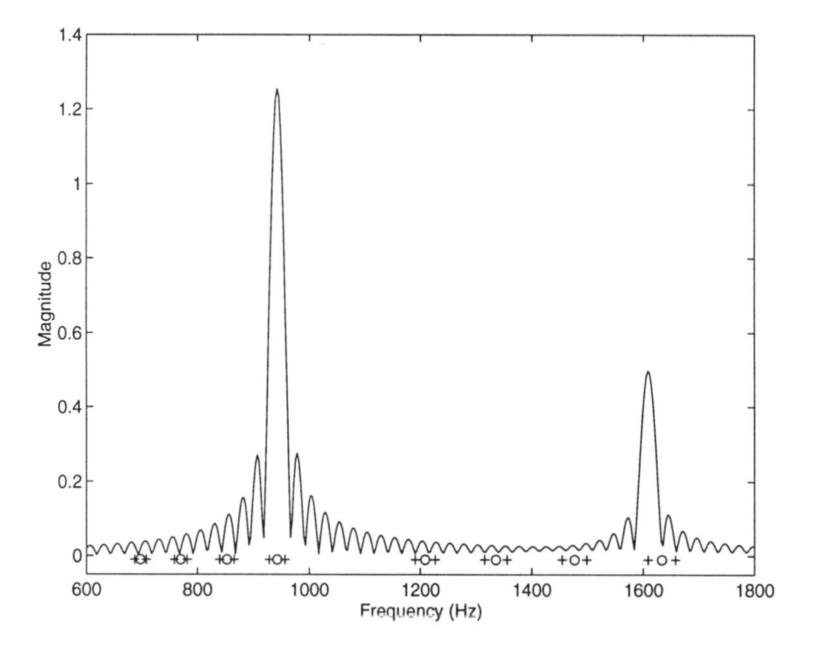

Figure 6.6. Effect of frequency tolerance on DTMF detection: This figure shows the magnitude of the Fourier transform of a DTMF signal with 320 samples, generated when the digit "*" is dialed. The eight DTMF frequencies can vary within the bands delimited by "+", about the center frequencies denoted by "o". Here the column tone is received at the lower end of the eighth band, and detection fails.

frequencies. The solid line represents the scenario when the DTMF signal is comprised of the row and column sinusoids at the operational frequencies. The dashed line corresponds to the addition of six more sinusoids with 30 dB lower amplitudes at the nonoperational frequencies. In the presence of these nonoperational tones, the resulting figure of merit may be lower or higher, depending on the window length N.

6.4.3 Performance Comparison

A performance comparison between four DTMF decoding algorithms is presented here. We determine the figure of merit F attained by each algorithm for a given number of input samples N. For each value of N, the worst-case figure of merit is determined by considering all the 16 DTMF digits, each under three twist conditions (no twist, maximum reverse twist, and maximum normal twist), and then taking the lowest value of F obtained. The DTMF signal is generated by adding two sinusoids at the operational row and column

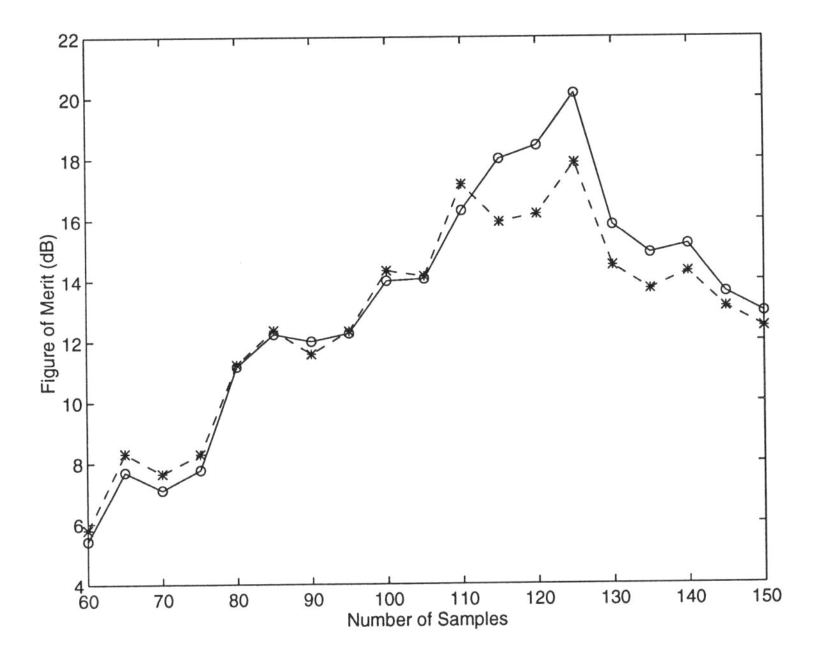

Figure 6.7. Effect of nonoperational tones on DTMF detection: Variation of the figure of merit with the number of samples in the DTMF signal, when the digit "1" is dialed. The solid line shows the case when sinusoids at the two operational frequencies are present in the signal. The dashed line shows the case when the signal also includes sinusoids with 30 dB lower amplitudes at the six nonoperational frequencies.

frequencies, and six sinusoids with 30 dB lower amplitudes at the remaining six DTMF frequencies.

We have performed simulations for *two cases*:

(1) The operational tones ω_r and ω_c are *fixed* at the standard DTMF frequencies.

(2) The operational frequencies *vary* within bands, ± 1.5 % about the standard frequencies.

In Case 2, we simulate the receiving conditions specified in the AT&T TouchTone standard by varying the two operational frequencies within tolerance bands, ± 1.5 % about the standard frequencies. Specifically, equidistant frequencies with a step size $|\Delta f| = 5$ Hz, are applied for each tolerance band. Thus, we have $5, 5, 7, 7, 9, 9, 9, 11$ possible choices of frequency in the eight frequency bands, respectively (since the widths of these bands are $20.9, 23.1, 25.6, 28.2, 36.3, 40.1, 44.3, 49$ Hz). For the various possible choices of operational DTMF frequencies, the worst-case figure of merit is computed.

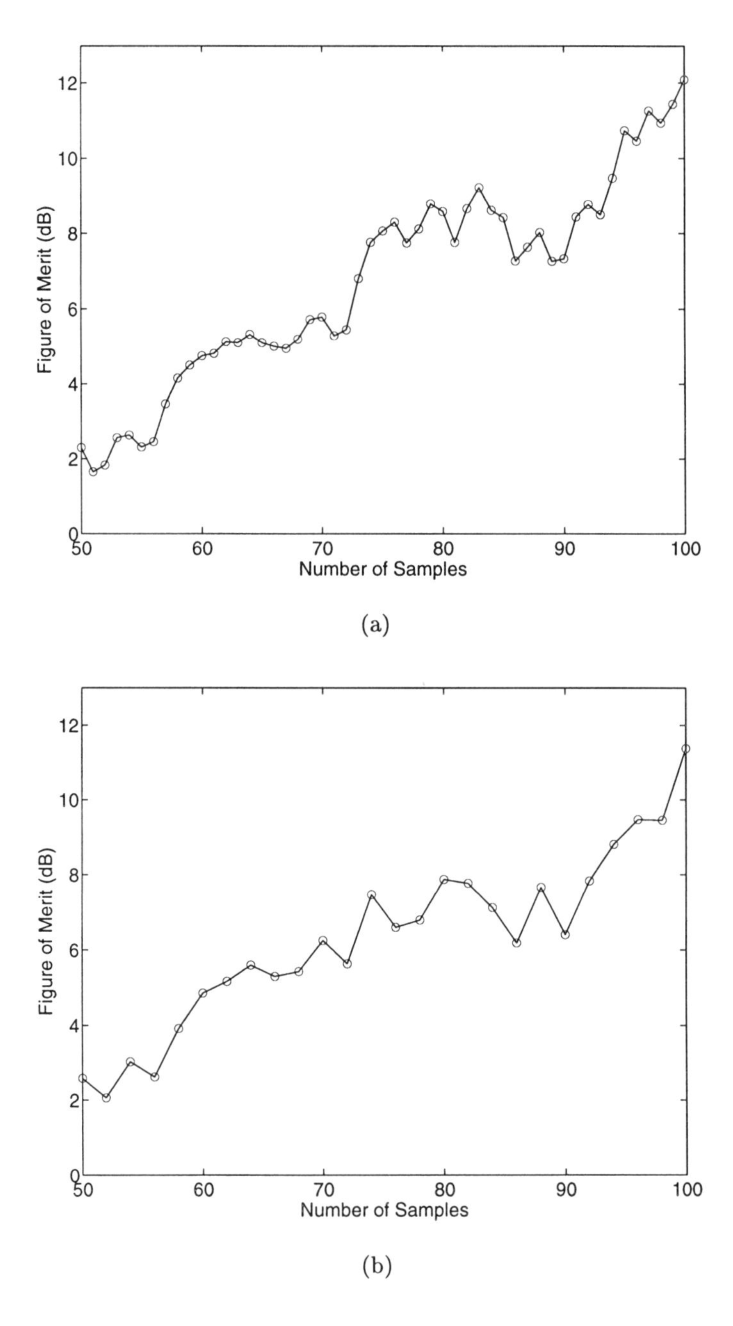

Figure 6.8. Variation of figure of merit with number of samples, for fixed operational frequencies. (a) NDFT. (b) SB-NDFT.

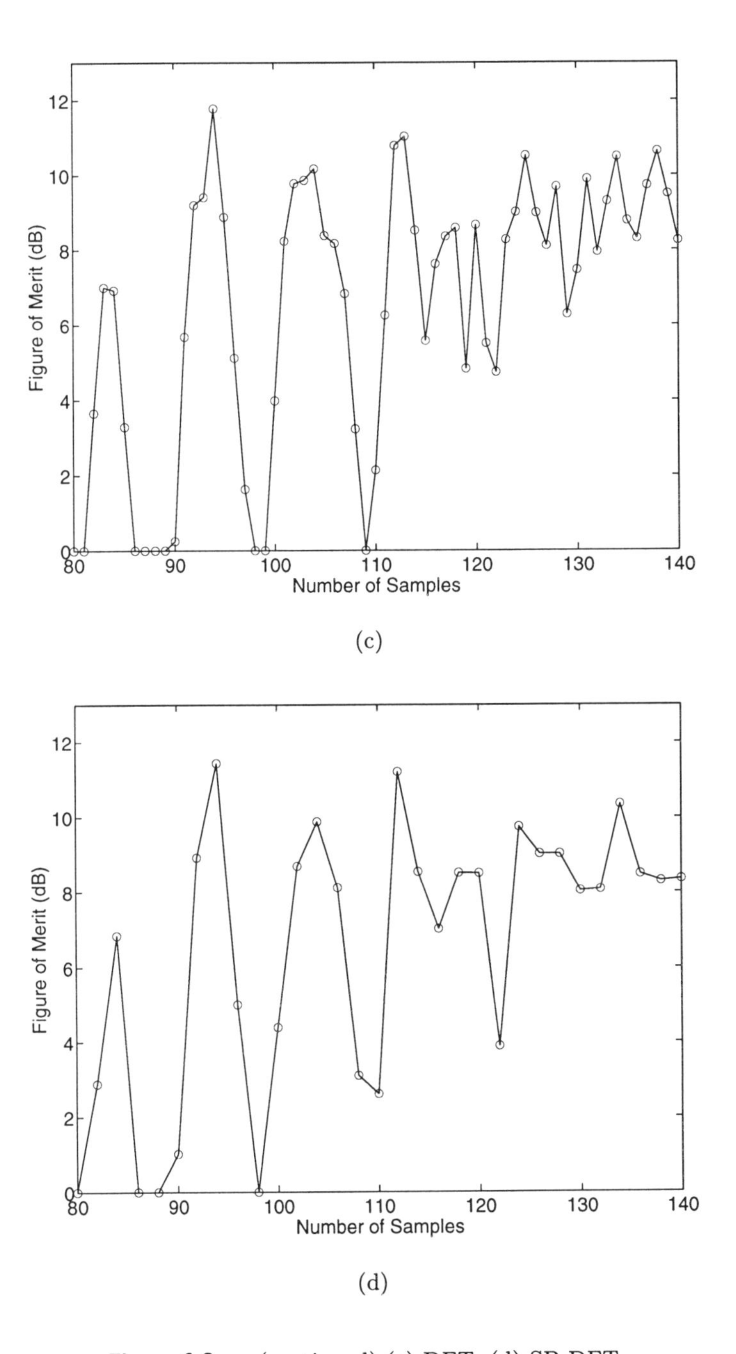

Figure 6.8. (continued) (c) DFT. (d) SB-DFT.

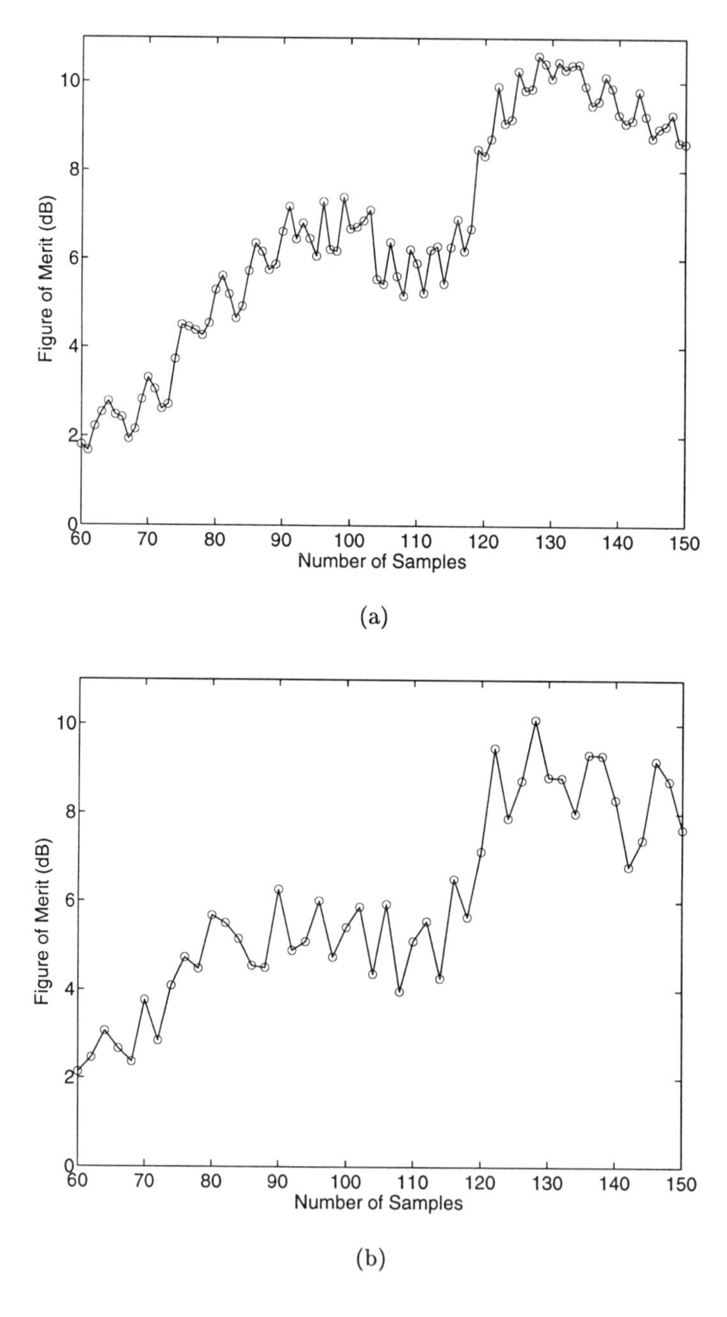

Figure 6.9. Variation of figure of merit with number of samples, for varying operational frequencies. (a) NDFT. (b) SB-NDFT.

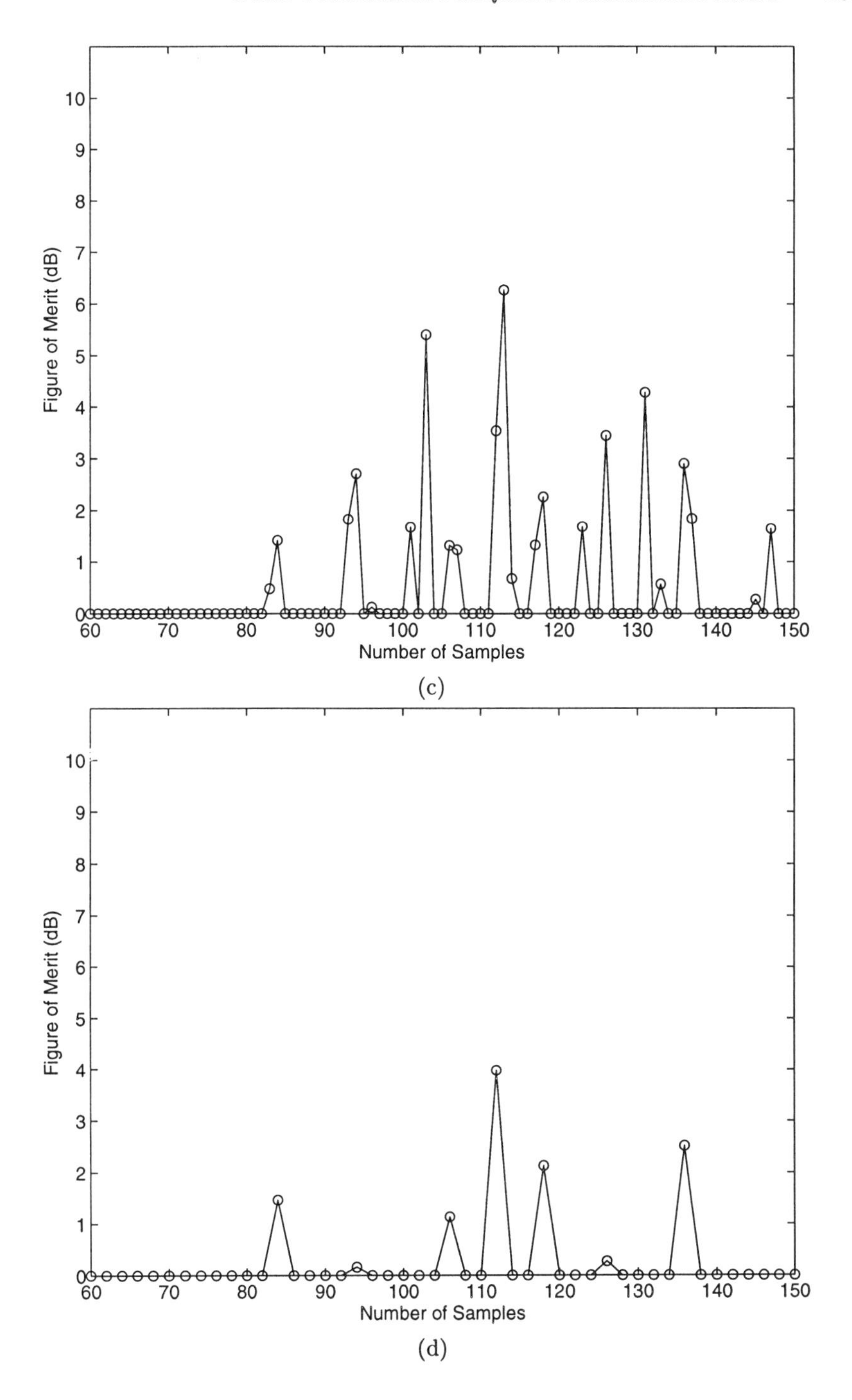

Figure 6.9. (continued) (c) DFT. (d) SB-DFT.

Table 6.2. Minimum number of samples required by DTMF decoding algorithms, to attain various figures of merit, with fixed operational frequencies.

Algorithm	Minimum number of samples required		
	$F = 3$ dB	$F = 6$ dB	$F = 9$ dB
NDFT	57	73	94
SB-NDFT	54	70	96
DFT	82	83	93
SB-DFT	84	84	94

Figs. 6.8 and 6.9 show the variation of the worst-case figure of merit with number of samples for Cases 1 and 2, respectively, using the four decoding algorithms—NDFT, SB-NDFT, DFT, and SB-DFT.

In view of the frequency tolerance requirements of the DTMF standard, Case 2 is of practical interest in choosing a decoding algorithm. The DFT and SB-DFT algorithms perform very poorly in this respect. The NDFT and SB-NDFT algorithms perform much better since the figures of merit attained by them in Case 2 are only slightly lower than those attained in Case 1. In these plots, a zero-valued figure of merit implies that the algorithm failed to detect one or more of the DTMF digits dialed. The DFT and SB-DFT algorithms fail for several intermediate values of N. This occurs due to the mismatch between the DFT sample frequencies and the DTMF standard frequencies, caused by the uniform frequency sampling used in the DFT.

We compare the four algorithms quantitatively by determining the *minimum number of samples* required to attain a certain figure of merit. This is shown in Tables 6.2 and 6.3 for Cases 1 and 2, respectively, for three values of the figure of merit—3 dB, 6 dB, and 9 dB.

As shown in Table 6.3, we cannot attain a high figure of merit using the DFT and SB-DFT algorithms, while complying with the frequency tolerance requirement. Finally, we determine the *total computation* required to detect the energies present in the input signal at the eight DTMF frequencies, using the minimum number of samples given in Table 6.3. The number of real multiplications M, and the number of real additions A, are shown in Table 6.4. Clearly, the SB-NDFT algorithm requires the *least computation* to attain a certain figure of merit. The gain is substantial when compared to other methods, including the one based on the NDFT.

Table 6.3. Minimum number of samples required by DTMF decoding algorithms to attain various figures of merit, with varying operational frequencies. The entry "–" indicates that the particular figure of merit is never attained.

Algorithm	Minimum number of samples required		
	$F = 3$ dB	$F = 6$ dB	$F = 9$ dB
NDFT	70	86	122
SB-NDFT	64	90	122
DFT	103	113	–
SB-DFT	112	–	–

Table 6.4. Computation required by DTMF decoding algorithms, to detect the energies present at the eight DTMF frequencies using the minimum number of samples. (M and A denote the number of real multiplications and real additions, respectively.)

Algorithm	$F = 3$ dB		$F = 6$ dB		$F = 9$ dB	
	M	A	M	A	M	A
NDFT	592	1136	720	1392	1008	1968
SB-NDFT	296	560	400	781	528	1053
DFT	856	1664	936	1824	–	–
SB-DFT	488	968	–	–	–	–

6.4.4 Detection of Second Harmonics

As discussed in Section 6.3, the SB-NDFT algorithm extracts the squared magnitudes of the DTMF tones by using the lower subband in a two-band decomposition. Fig. 6.10 shows the spectrum that results due to this approximation. The DTMF tones are assumed to have unity magnitude, as shown in Fig. 6.10(a). The magnitude spectrum obtained by retaining only the lower subband is shown in Fig. 6.10(b); this corresponds to the first term of Eq. (6.5), $(1 + z_k^{-1})\, G_L(z_k^2)$. The second set of eight tones (symmetric about $f_s/4 = 2000$ Hz and occurring in reverse order) arises due to the downsampling in the preprocessing stage of the SB-NDFT algorithm. Fig. 6.10(c) shows the magnitude spectrum obtained by retaining only the higher subband, i.e., the second term of Eq. (6.5), $(1 - z_k^{-1})\, G_H(z_k^2)$.

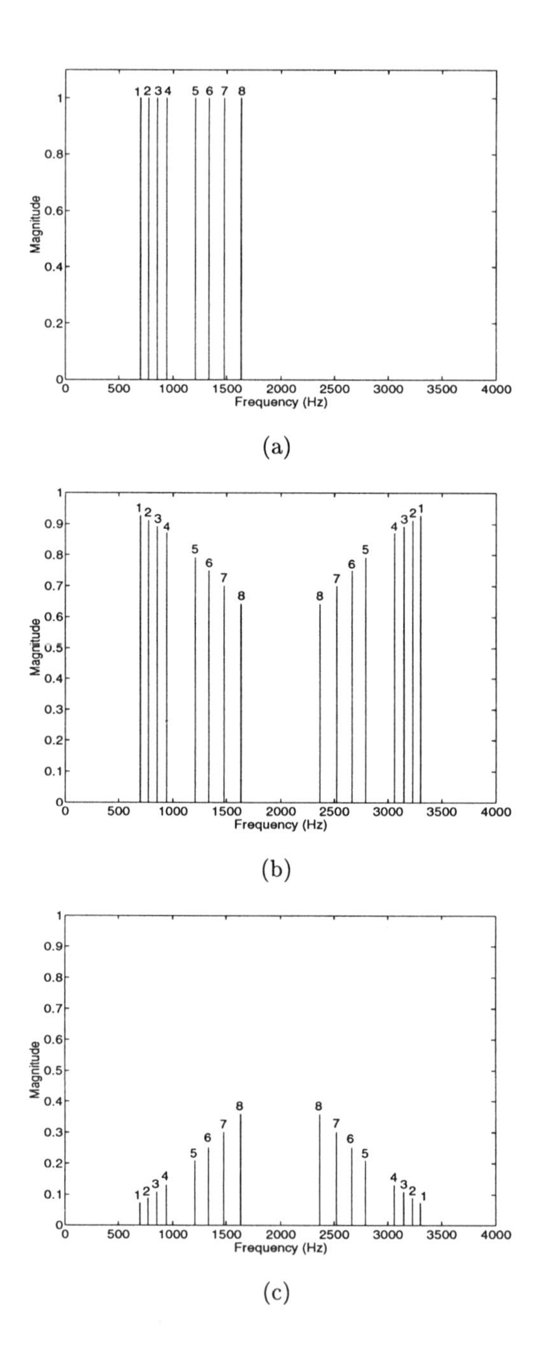

Figure 6.10. Magnitude spectrum of DTMF tones obtained by SB-NDFT approximation. (a) Original spectrum of DTMF tones. (b) Spectrum obtained by retaining only the lower subband. (c) Spectrum obtained by retaining only the higher subband.

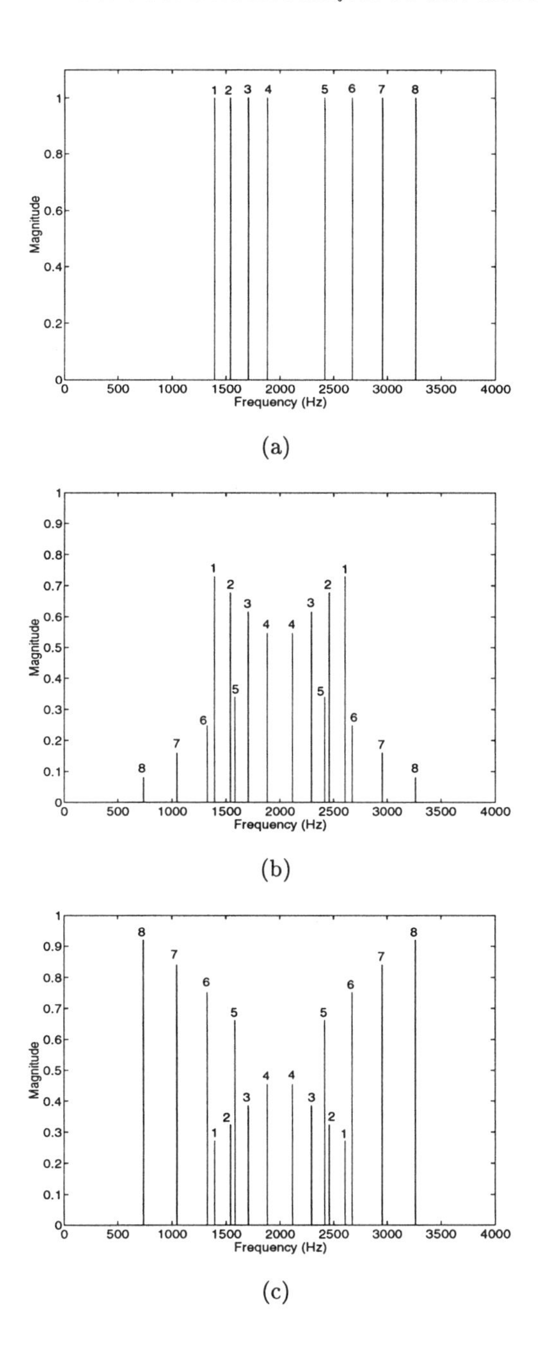

Figure 6.11. Magnitude spectrum of DTMF second harmonics obtained by SB-NDFT approximation. (a) Original spectrum of second harmonics of DTMF tones. (b) Spectrum obtained by retaining only the lower subband. (c) Spectrum obtained by retaining only the higher subband.

Since the second harmonics of the DTMF tones are usually detected to guard against digit simulation by speech, we consider the magnitude spectrum of the second harmonics in Fig. 6.11. The results of retaining the lower and higher subbands are illustrated in Figs. 6.11(b) and (c), respectively. Aliasing of the second harmonics and their symmetric frequencies occurs because of downsampling. Note that the second harmonics have more energy in the higher subband. Therefore, we can compute their magnitudes from this subband. The thresholds for digit validation have to be determined accordingly. In a DSP implementation of this algorithm, its robustness in rejecting speech can be tested by using the Mitel test tape [Mitel Semiconductor] which contains digit simulation speech recordings.

6.5 SUMMARY

In this chapter, we proposed a new algorithm for DTMF signal decoding based on the subband NDFT. This algorithm utilizes the fact that DTMF frequencies reside in the low-frequency part of the telephone bandwidth. We obtained a good approximation to the energies present at the DTMF frequencies by discarding calculations in the high-frequency band of a two-band decomposition. Extensive simulations were carried out based on the AT&T TouchTone standard, to compare the performances of four DTMF decoding algorithms based on the NDFT, SB-NDFT, DFT, and SB-DFT. We found that the SB-NDFT algorithm requires much lower computation than the others to attain a desired level of performance. This result is significant because, when implemented on a DSP, the proposed algorithm can lead to an increase in the number of DTMF channels that can be decoded simultaneously in a multi-channel environment.

7 CONCLUSIONS

The concept of the nonuniform discrete Fourier transform (NDFT) was introduced to provide a generalized approach for nonuniform sampling in the frequency domain. The NDFT of a finite-length sequence corresponds to sampling its z-transform at arbitrarily chosen points in the z-plane. The NDFT reduces to the discrete Fourier transform (DFT) when these sampling points are located on the unit circle at equally spaced angles. The flexibility in sampling offered by the NDFT leads to a variable spectral resolution which can be controlled by the user. This is important in many signal processing applications, since most signals and systems tend to have their energies distributed unevenly in different regions of the spectrum. We developed applications of the NDFT in the areas of spectral analysis, 1-D and 2-D FIR filter design, antenna pattern synthesis with prescribed nulls, and dual-tone multi-frequency tone detection.

After defining the NDFT, we showed that the inverse NDFT exists provided the sampling points are distinct. The problem of computing the inverse NDFT is mathematically equivalent to the problem of polynomial interpolation. We studied some relevant properties of the NDFT. The Goertzel algorithm can be used for computing the NDFT. However, algorithms as efficient as the FFT cannot be derived because there is no periodicity in the generalized sampling structure provided by the NDFT. We outlined the subband NDFT, which is a method for computing the NDFT based on a subband decomposition of the input sequence. This leads to a fast, approximate computation of the NDFT for signals which have their energy concentrated in a few bands of the spectrum.

The concept of the NDFT was also extended to two dimensions. Some special cases of the 2-D NDFT were mentioned, including the 2-D DFT.

We first developed an application of the NDFT in 1-D FIR filter design. A new filter design method was proposed using nonuniform frequency sampling. In this method, we represent the desired frequency response by analytic functions, one for each band. These functions are based on Chebyshev polynomials and involve several parameters which are given by closed-form expressions depending on filter specifications such as band edges and peak ripple. The desired response is then sampled at the extremal frequencies. An inverse NDFT of these samples yields the filter coefficients. The performance of this method is superior compared with earlier design methods based on uniform frequency sampling. The resulting filters are nearly equal to optimal (minimax) filters with the same design specifications. In addition, the design times are much lower than those required by the Parks-McClellan algorithm. The proposed nonuniform choice of extrema can also serve as a good starting point for the Parks-McClellan algorithm, if exactly optimal filters are to be designed. Simulations show that this choice decreases the number of iterations required, as well as the design time.

The 2-D NDFT was used to develop a nonuniform frequency sampling technique for designing 2-D FIR filters. A suitable cross-section of the desired 2-D frequency response is approximated by equiripple 1-D analytic functions based on Chebyshev polynomials. Samples are then placed on contours that have the desired 2-D shape and pass through the extrema of this cross-section. Earlier nonuniform frequency sampling design methods did not lay down clear guidelines for the choice of sample values and locations. Our method produces nonseparable 2-D filters with good passband shapes and low peak ripples. The results are significant owing to the lack of a practical, reliable algorithm to design optimal 2-D filters.

We presented another application of the NDFT for synthesizing antenna patterns with prescribed nulls. Such patterns are often used to suppress interfering signals incident on an array antenna from certain specific directions. Our synthesis technique is based on nonuniform sampling in the angular domain. The original array pattern is first represented by samples located at the extrema in the angular domain. The nulls are then imposed by placing zero-valued samples on the pattern at the specified angles. An inverse NDFT of the samples yields the weights of the constrained pattern. This process is repeated with the new pattern in successive iterations, without disturbing the zero-valued samples. The proposed algorithm is very effective in producing patterns with deep nulls within only two or three iterations.

Finally, we proposed an efficient algorithm for decoding dual-tone multi-frequency (DTMF) signals, based on the subband NDFT (SB-NDFT). As specified in a CCITT standard, the DTMF signalling frequencies occupy the low-frequency part of the telephone bandwidth. By utilizing this information, we developed an efficient algorithm to approximately compute the energies present at the DTMF frequencies. In this algorithm, we decompose the input signal into two subbands, and retain only the lower band to compute the energies using the modified Goertzel algorithm. Simulations were performed, based on the AT&T TouchTone standard, to evaluate the performance of this algorithm in comparison with other decoding algorithms based on the DFT, SB-DFT, and NDFT. Our results demonstrate that the SB-NDFT algorithm achieves substantial savings in computation to attain a desired level of performance.

We have established that the NDFT can serve as an important tool in many signal processing applications. In all the applications considered here, the NDFT-based algorithms performed much better than those based on the DFT. This demonstrates the effectiveness of nonuniform frequency sampling, provided the freedom available in locating the samples is utilized properly.

References

Aly, S. and Fahmy, M. (1981). Symmetry in two-dimensional rectangularly sampled digital filters. *IEEE Trans. Acoust., Speech, Signal Processing*, ASSP-29:794–805.

Angelides, E. (1994). A novel method for modeling 2-D FIR digital filters in frequency domain with nonuniform samples. *IEEE Trans. Circuits Syst. II: Analog and Digital Signal Processing*, 41:482–486.

Angelides, E. and Diamessis, J. (1994). A novel method for designing FIR digital filters with nonuniform frequency samples. *IEEE Trans. Acoust., Speech, Signal Processing*, 42:259–267.

Ansari, R. and Reddy, M. (1987). Efficient IIR and FIR fan filters. *IEEE Trans. Circuits Syst.*, CAS-34:941–945.

Atkinson, K. (1978). *An Introduction to Numerical Analysis.* John Wiley & Sons, New York.

Bagchi, S. (1994). *The nonuniform discrete Fourier transform and its applications in signal processing.* Ph.D. dissertation, University of California at Santa Barbara.

Bagchi, S. and Mitra, S. (1995a). An efficient algorithm for DTMF decoding using the subband NDFT. In *Proc. IEEE Int. Symp. on Circuits and Syst.*, volume 3, pages 1936–1939, Seattle, WA.

Bagchi, S. and Mitra, S. (1995b). Nonseparable 2-D FIR filter design using nonuniform frequency sampling. In *Proc. IS&T/SPIE Symposium on Electronic Imaging: Image and Video Processing III*, pages 104–115, San Jose, CA.

Bagchi, S. and Mitra, S. (1996a). The nonuniform discrete Fourier transform and its applications in filter design: Part I—1-D. *IEEE Trans. Circuits Syst. II: Analog and Digital Signal Processing*, 43:422–433.

Bagchi, S. and Mitra, S. (1996b). The nonuniform discrete Fourier transform and its applications in filter design: Part II—2-D. *IEEE Trans. Circuits Syst. II: Analog and Digital Signal Processing*, 43:434–444.

Bagchi, S. and Mitra, S. (1997). Efficient robust DTMF decoding using the subband NDFT. *Signal Processing*, 56:255–267.

Bamberger, R. and Smith, M. (1992). A filter bank for the directional decomposition of images: Theory and design. *IEEE Trans. Signal Processing*, 40:882–893.

Blahut, R. (1985). *Fast Algorithms for Digital Signal Processing.* Addison-Wesley, Reading, MA.

Carroll, C. and Kumar, B. V. (1989). Pattern synthesis using Fourier transforms. *Optical Engineering*, 28:1203–1210.

Chen, T. and Vaidyanathan, P. (1993). Multidimensional multirate filters and filter banks derived from one-dimensional filters. *IEEE Trans. Signal Processing*, 41:1749–1765.

Collin, R. and Zucker, F. (1969). *Antenna Theory, Part I.* McGraw-Hill, New York.

Cooley, J. and Tukey, J. (1965). An algorithm for the machine computation of complex Fourier series. *Mathematics of Computation*, 19:297–301.

Davis, P. (1975). *Interpolation and Approximation.* Dover Publications, New York.

Diamessis, J., Therrien, C., and Rozwood, W. (1987). Design of 2-D FIR filters with nonuniform frequency samples. In *Proc. IEEE Int. Conf. on Acoust., Speech, Signal Processing*, volume 3, pages 1665–1668, Dallas, TX.

Dolph, C. (1946). A current distribution for broadside arrays which optimizes the relationship between beam width and side-lobe level. *Proc. IRE*, 34:335–348.

DuHamel, R. (1953). Optimum patterns for endfire arrays. *Proc. IRE*, 41:652–659.

Gay, S., Hartung, J., and Smith, G. (1989). Algorithms for multi-channel DTMF detection for the WE DSP32 family. In *Proc. IEEE Int. Conf. on Acoust., Speech, Signal Processing*, volume 2, pages 1134–1137, Glasgow, UK.

Goertzel, G. (1958). An algorithm for the evaluation of finite trigonometric series. *American Math. Monthly*, 65:34–35.

Gold, B. and Jordan, Jr., K. (1969). A direct search procedure for designing finite duration impulse response filters. *IEEE Trans. Audio Electroacoust.*, AU-17:33–36.

Gold, B. and Rader, C., editors (1969). *Digital processing of Signals.* McGraw-Hill, New York.

Golub, G. and Loan, C. V. (1983). *Matrix Computations.* The John Hopkins University Press, Baltimore.

Harris, D. and Mersereau, R. (1977). A comparison of algorithms for minimax design of two-dimensional linear phase FIR digital filters. *IEEE Trans. Acoust., Speech, Signal Processing*, ASSP-25:492–500.

Hazra, S. and Reddy, M. (1986). Design of circularly symmetric low-pass two-dimensional FIR digital filters using transformation. *IEEE Trans. Circuits Syst.*, CAS-33:1022–1026.

Hu, J. and Rabiner, L. (1972). Design techniques for two-dimensional filters. *IEEE Trans. Audio Electroacoust.*, AU-20:249–257.

Huang, T., Burnett, J., and Deczky, A. (1975). The importance of phase in image processing filters. *IEEE Trans. Acoust., Speech, Signal Processing*, ASSP-23:529–542.

International Telecommunication Union (1984). CCITT Red Book. Volume VI, Fascicle VI.1.

Jarske, P., Neuvo, Y., and Mitra, S. (1988). Improved frequency sampling FIR filter design. In *Proc. EUSIPCO '88, Fourth European Signal Processing Conf.*, volume 2, pages 691–694, Grenoble, France.

Kaiser, J. (1974). Nonrecursive digital filter design using the I_0-sinh window function. In *Proc. IEEE Int. Symp. on Circuits and Syst.*, pages 20–23, San Francisco, CA.

Kamp, Y. and Thiran, J. (1975). Chebyshev approximation for two-dimensional nonrecursive digital filters. *IEEE Trans. Circuits Syst.*, CAS-22:208–218.

Leger, A., Omachi, T., and Wallace, G. (1991). JPEG still picture compression algorithm. *Optical Engineering*, 30:947–954.

Lightstone, M., Mitra, S., Lin, I., Bagchi, S., Jarske, P., and Neuvo, Y. (1994). Efficient frequency-sampling design of one-and two-dimensional FIR filters using structural subband decomposition. *IEEE Trans. Circuits Syst. II: Analog and Digital Signal Processing*, 41:189–201.

Lim, J. (1990). *Two-Dimensional Signal and Image Processing.* Prentice-Hall, Englewood Cliffs, N.J.

Lodge, J. and Fahmy, M. (1980). An efficient l_p optimization technique for the design of two-dimensional linear-phase FIR digital filters. *IEEE Trans. Acoust., Speech, Signal Processing*, ASSP-28:308–313.

Marr, A., editor (1990). *Digital Signal Processing Applications Using the ADSP-2100 Family.* Prentice-Hall, Englewood Cliffs, N.J. Chapter 14.

McClellan, J. (1973). The design of two-dimensional digital filters by transformation. In *Proc. 7th Ann. Princeton Conf. Inform. Sci. and Syst.*, pages 247–251, Princeton, NJ.

McClellan, J., Parks, T., and Rabiner, L. (1973). A computer program for designing optimum FIR linear phase digital filters. *IEEE Trans. Audio Electroacoust.*, AU-21:506–526.

Mersereau, R., Mecklenbraäuker, W., and T.F. Ouatieri, J. (1976). Mcclellan transformations for two-dimensional digital filtering: I – Design. *IEEE Trans. Circuits Syst.*, CAS-23:405–414.

Mintzer, F. (1982). On half-band, third-band and Nth band FIR filters and their design. *IEEE Trans. Acoust., Speech, Signal Processing*, ASSP-30:734–738.

Mitel Semiconductor. Tone Receiver Test Cassette # CM7291, Mitel Technical Data Manual.

Mitra, S. (1998). *Digital Signal Processing: A Computer-Based Approach.* McGraw-Hill, New York, N.Y.

Mitra, S., Chakrabarti, S., and Abreu, E. (1992). Nonuniform discrete Fourier transform and its applications in signal processing. In *Proc. EUSIPCO '92,*

Sixth European Signal Processing Conf., volume 2, pages 909–912, Brussels, Belgium.

Mitra, S., Shentov, O., and Petraglia, M. (1990). A method for fast approximate computation of discrete-time transforms. In *Proc. IEEE Int. Conf. on Acoust., Speech, Signal Processing*, volume 4, pages 2025–2028, Albuquerque, NM.

Mock, P. (1985). Add DTMF generation and decoding to DSP μP designs. *EDN*, 30:205–220.

Oppenheim, A. and Johnson, D. (1971). Computation of spectra with unequal resolution using the fast Fourier transform. *Proc. IEEE*, 59:299–301.

Oppenheim, A. and Schafer, R. (1989). *Discrete-Time Signal Processing.* Prentice-Hall, Englewood Cliffs, N.J.

Pratt, W. (1991). *Digital Image Processing.* John Wiley & Sons, New York.

Pritchard, R. (1955). Discussion on optimum patterns for endfire arrays. *IRE Trans. Antennas Propagat.*, AP-3:40–43.

Rabiner, L., Gold, B., and McGonegal, C. (1970). An approach to the approximation problem for nonrecursive digital filters. *IEEE Trans. Audio Electroacoust.*, AU-18:83–106.

Rabiner, L., Schafer, R., and Rader, C. (1969). The chirp z-transform algorithm. *IEEE Trans. Audio Electroacoust.*, AU-17:86–92.

Regalia, P. and Mitra, S. (1989). Kronecker products, unitary matrices, and signal processing applications. *SIAM Review*, 31:586–613.

Riblet, H. (1947). Discussion on "A current distribution for broadside arrays which optimizes the relationship between beam width and side-lobe level". *Proc. IRE*, 35:489–492.

Rozwood, W., Therrien, C., and Lim, J. (1991). Design of 2-D FIR filters by nonuniform frequency sampling. *IEEE Trans. Acoust., Speech, Signal Processing*, ASSP-39:2508–2514.

Schüssler, W. (1972). On structures for nonrecursive digital filters. *Arch. Elek. Übertragung*, 26:255–258.

Shentov, O., Hossen, A., Mitra, S., and Heute, U. (1991). Subband DFT—interpretation, accuracy, and computational complexity. In *Twenty-Fifth Asilomar Conf. on Signals, Systems, and Computers*, volume 1, pages 95–100, Pacific Grove, CA.

Shentov, O. and Mitra, S. (1991). A simple method for power spectral estimation using subband decomposition. In *Proc. IEEE Int. Conf. on Acoust., Speech, Signal Processing*, volume 5, pages 3153–3156, Toronto, Canada.

Shentov, O., Mitra, S., Heute, U., and Hossen, A. (1995). Subband DFT - Part I: Definition, interpretation and implementation. *Signal Processing*, 41:261–277.

Siohan, P. (1991). 2-D FIR filter design for sampling structure conversion. *IEEE Trans. Circuits Syst. for Video Technology*, 1:337–350.

Steykal, H. (1982). Synthesis of antenna patterns with prescribed nulls. *IEEE Trans. Antennas Propagat.*, AP-30:273–279.

Steykal, H. (1983). Wide-band nulling performance versus number of pattern constraints for an array antenna. *IEEE Trans. Antennas Propagat.*, AP-31:159–163.

Steykal, H., Shore, R., and Haupt, R. (1986). Methods for null control and their effects on the radiation pattern. *IEEE Trans. Antennas Propagat.*, AP-34:404–409.

Tonge, G. (1981). The sampling of television images. Experimental and Development Rep. 112/81, Independent Broadcasting Authority.

Yoshida, T., Nishihara, A., and Fujii, N. (1990). A design method of 2-D maximally flat diamond-shaped half-band FIR filters. *Trans. IEICE (The Institute of Electronics, Information and Communication Engineers)*, E 73:901–907.

Zakhor, A. and Alvstad, G. (1992). Two-dimensional polynomial interpolation from nonuniform samples. *IEEE Trans. Acoust., Speech, Signal Processing*, ASSP-40:169–180.

Index